Einstein's Destruction of Physics

Peter Šujak

ISBN 978-1-387-57644-9

Einstein's Destruction of Physics
Author: Peter Šujak
Pages: 210
Language: English
First published: August 2017

Second edition

Published and printed by Lulu Press, Inc., February 2018
627 Davis Drive, Suite 300, Morrisville, NC 27560, USA
www.lulu.com

ISBN   978-1-387-57644-9

Acknowledgements

This project began in my mind as a result of discussions with many colleagues in the 1990's at laboratories in the Nuclear Power Plant Research Institute located in the Slovak town of Trnava. In this Institute I obtained my first job in 1981 after I earned my Masters degrees in Experimental Nuclear Physics at the Comenius University in Bratislava, the Capital city of Slovakia. In 1982 I earned a doctorate there in Nuclear Physics as well.

I also much appreciate my interaction with many colleagues at the Headquarters of Radiation Monitoring Centre in Prague, where I worked from 1987 until 1990.

I am most grateful to various authorities and professors at the Faculty of Mathematics and Physics, Charles University in Prague, who enabled me to attend my extraordinary recharging study in the ground physical platform of contemporary fundamental physics from 2008 until 2010.

Further thanks to my friend, mathematician Radek Hořeňovský, for his uncountable and patient discussions during the last 10 years about these many physical problems.

I also wish to thank Michaela and Jim Freeman for their multiple proofreading of this book.

The support and understanding of my mother, sister and nephew mean more to me than can ever be expressed in words. I hope my loved daughter understands why I missed the free time we ought to have spent together.

Preface

This book is intended for anyone who is interested in a real physical image and order of the physical world surrounding us. It represents a concise summary of the abolishment of principles of formal logic in relativity and quantum mechanics.

Einstein's destruction of physics and scientific principles is documented here. The real substance of gravity and inertial forces is debunked. The substance of mass is recovered and the nature of time in physics is revealed. The reality of the double-slit experiment is revealed. The physical reality of mysteries of space, time, frequencies, wavelengths and wave functions in Relativity and Quantum mechanics is exposed. It shows that Quarks and Higgs bosons do not exist.

I document that, for the past four hundred years, there is no distinguished physicist who would not have recognized that without existence of the ether it is not possible to explain the physical world around us. It show that Einstein, after 1916, came in with his rediscovery of the ether and he subsequently became a more enthusiastic advocate of the proven existence of the ether than supporters of the ether before 1905.

The book explains that all elementary particles, all rigid matter and all force fields in the Universe are created from compression of ether. Filling the space of the Universe with swirling ether is all that is required for the self-evolution of the Universe.

My book provides an overview of the opposition of physicists against the mainstream physical image of the world over the past hundred years. It documents the basic historical, philosophical and physical reasons for denial of the main physical theories of the 20th century.

The aim of my book is to return physics from its way of metaphysics in the 20th century on the way of the physical reality in the 21st century. Several equations of relativity and quantum mechanics are included in chapters 4 and 6, written in their most simplified form and manifesting their physical concepts to the general public. The book is the result of 10 years intense research work in the ground physical platform of contemporary fundamental physics.

Prague   31 May 2017   Šujak Peter

Preface to the second edition

The second edition of this book was augmented by twenty pages compared to its first edition. Mainly chapters 2., 3., and 4. were augmented by around five pages each. After this augmentation it appears that the argumentation about the unacceptability of the ill-founded physical theories of the 20th century represents a compact corpus.

Prague   25 January 2018   Šujak Peter

Contents

1. Introduction

Since the appearance of thinking man on Earth, he has attempted to find an answer to the question of the meaning of the existence of his own life. The answer to this question is connected with the answer to the question of what is the meaning of life itself and of the universe, as well as what is the image and order of the surrounding physical world and man's position in it.

The ruling power structures reduced the misery of man with the answer to this question and mainly offered answers that ensured his loyalty. Physics and other sciences have always been the tools for gaining the ascendancy and domination of one group of people over others. The physical picture of the world has always been a special tool used by the ruling power structures to maintain control over their own nations and states.

Therefore, from the ancient times of shamans, pharaohs and great priests of various religions up to today, the officially established order of the universe has been a single untouchable truth. Opposition or the questioning of this truth was perceived as an attack on the establishment of governmental power structures.

These shamans, pharaohs and great priests of various religions declared themselves the intermediators between God and man. They brought to man their 'God laws' of creation and the existence of this world along with God laws of their percepts and commandments, addressed to man. Each man must abide by these laws in order to achieve a promised paradise and to avoid specific punishments and havoc for disobedience. Ensuring the obedience of man is the entire point of these God given laws.

The veracity of this order of the universe was a minor issue for the ruling powers, as is also the case with those physical theories of the 20th century.

The purpose of these physical theories espoused by theoretical physicists and produced during the 20th century (special and general relativity, quantum mechanics, force fields theories using force-mediating particles, the standard model of quark theory, the Big Bang and the Higgs boson theories) was to conceal the existence of physical fields as an actual material media. The force action of these media is replaced in these theories by the mathematics of kinematical non-material quantities.

In fact, these kinematical quantities describe the physical material world, which works on the principle of balance and changes of pressure of the densities of the material bodies and of the material force fields surrounding these bodies. Without the recognition of the existence of physical fields as actual physical substance, all mathematical descriptions of physical reality are merely the kinematical numbers of the ratios of the lengths and times on paper.

Even the easiest physical processes, such as the collision of two bodies, cannot be reasonably explained by a thinking person without the existence of the force fields associated with these bodies.

Newton... "*without the Mediation of anything else, by and through which their Action and Force may be conveyed from one to another, is to me so great an Absurdity that I believe no Man who has in philosophical Matters a competent Faculty of thinking can ever fall into it*".

Maxwell..."*In fact, whenever energy is transmitted from one body to another, there must be a medium or substance in which the energy exists. ...all theories lead to the conception of a medium in which the propagation takes place*".

Poincaré in 1909: "*In this new concept, the constant mass of matter disappeared. The ether alone, and not matter anymore, was inertial. Only the ether opposed to a resistance to motion, thus one could say: there was no matter, there were only holes in the ether*" [50].

Einstein after 1920, as detailed in chapters 2 and 3 ... "*The mechanical ether, designated by Newton as 'absolute space', must therefore be considered by us as a physical reality. According to our present conceptions the elementary particles of matter are also, in their essence, nothing else than condensations of the electromagnetic field. ... now it appears that ether will have to be regarded as a primary thing and that matter is derived from it".*

Under the supervision of the ruling power structures and the mass information media, these physically deformed theories of the 20th century are untouchable truths of the order of the universe, as was the case in the past in Ptolemy's geocentric mathematical description of the universe-an untouchable truth for 1500 years.

Special relativity (STR) in 1905 removed from the physics the ether as a real physical media discovered by many generations of physicists, at hundred year intervals, exploring the phenomena of electricity and magnetism during the 19th century. General relativity (GTR) in 1915 eliminated the ether from physics as a real physical media of the gravitational field discovered by many generations of physicists surrounding Newton. Although Riemann, whose differential geometry of curved space became the basic language of the GTR, asserts to us that space without material content filling it is nothing more than a three-dimensional manifold devoid of all form, this basic fundamental of Riemann is concealed in GTR.

Big Bang removed the media from the Intergalactic Space (ether, dark energy) about which Hubble (until the end of his life) assumed that it may be the main cause of the observed red shift in the spectra of galaxies. Quantum mechanics (QM) and the theory of Higgs boson removes the ether from physics, in quantum physics known as initial medium at creation and the annihilation of matter particles from and into electromagnetic waves.

Although astronomers bring daily evidence concerning the birth of millions stars yearly, as well as new galaxies from the shrinking of interstellar gas and dark energy, which is contrary to the expansion and cooling of the universe, the Big Bang expanding theory is forced upon the public as the only possible explanation for the creation of the universe. Although the observed redshift of spectra of galaxies can be physically explained in at least five possible ways, only this one, as the sole and irrefutable evidence of the expansion of the universe, is forced upon the public.

From the measurements of observed redshift of spectra of a Galaxy it cannot be distinguished from the kinematic point of view if the Galaxy is moving in a straight line away from us or is moving in any various direction up to a right angle from this straight line or moves along a circular path around us, but this fact is fully ignored in the Big Bang theory.

Dayton Miller after 30 years of conducting Michelson-Morley interferometric experiments, in 1933 fully determined the absolute motion and the speed of the earth in its surrounding ether and Einstein in 1913 came to declare that his *"special theory of relativity is only an approximation to the reality of changing light velocity"*.

These facts are concealed over the next 100 years and the validity of STR is forced upon both public and students by referring only to Einstein's 1905 paper and the purported null result of Michelson-Morley's experiment from 1887.

Einstein, after 1916, becomes a more enthusiastic advocate of the proven existence of the ether than supporters of the ether before the year 1905. Although he proclaimed in 1924 that *"the mechanical ether designated by Newton as 'absolute space' must therefore be considered by us as a physical reality"*, this fact was also concealed by the ideological power structures of the world for the next 100 years. Einstein's mystery of nonmaterial space time is daily forced upon the public and the wider physical community, referring only to Einstein's paper from 1915 although in 1920 he declared that *"The theory of space (geometry) and time no longer represents intrinsic physics"*

Today we know that, by annihilation of protons with antiprotons, we produce electromagnetic radiation and vice versa that by waving a magnet around a wire we produce electromagnetic waves that are able to produce protons. This experimental fact is the full evidence of the existence of the ether as a real physical substance. Yet current physics says that all matter was created solely from the fictitious mathematical construction of Higgs boson.

2. The Development of a Man's View of his Surrounding Physical World

Philosophers and astronomers of the culminant era (around 500-200 BC) during the development of the Greek philosophy of nature (Pythagoras, Democritus, and many others) were convinced [1, p. 511] that the Sun is in the center of the known universe. They further thought that under the stars drifting in the sky through the ether on a rotating sphere, six planets orbit around the Sun and that the Earth, as one of these planets, also orbits around the Sun in an annual cycle and that the Earth rotates daily around its own axis. They had numbered five comets and were convinced that comets circulate like planets around the Sun on a very eccentric orbit.

They were convinced that the Milky Way was made up of individual stars and had named the center of the Milky Way Nebula. They were convinced that matter was composed of the smallest indivisible particles of atoms. (N.B. For Greek atom today would be called our further indivisible proton).

The Greek astronomer and mathematician Aristarchus of Samos (c. 310 – c. 230 BC) is documented as the first to present explicit arguments for a heliocentric model of the Solar System, placing the Sun, not the Earth, at the center of the known universe. About 260 BC, in his work '*On the Sizes and Distances of the Sun and Moon*' he used trigonometry to estimate the size of the Moon and measured its distance by the Earth's shadow during a lunar eclipse. He described the Earth as rotating daily on its axis and revolving annually about the Sun in a circular orbit, while the Sun remained stationary like the stars [Wikipedia].

Archimedes of Syracuse (c. 287 – c. 212 BC) constructed a mechanism which showed the motion of Earth and five planets around the Sun and the Moon around the Earth. Cicero also mentioned that similar mechanisms were designed by Thales of Miletus (c. 624 – c. 546 BC) [Wikipedia].

From continuing recognition [2] of the artifact Antikythera clockwork mechanism, unearthed in 1900 – 1901 and dated to 200–100 BC, we know that ancient Greeks mastered the computing of eclipses of the Sun and the Moon, as well as the positions of stars in the sky and the position of the five planets in their orbit. They were even able to include a different velocity along the elliptical motion of the Moon around the Earth in this computing and obviously also a different velocity along the elliptical motion of the Earth around the Sun. The longest included period in this computing was the Callippic cycle (proposed by Callippus in 330 BC) that was explained by Hipparchus later as he fully discovered (127 BC) the precession of the Earth's axis in the period of 26,000 years.

Greek theoretical astronomy was based on the experimental observations of their predecessors in the previous thousands of years-- the Sumerians, Egyptians, Babylonians, Chaldeans and Babylonian star catalogues appearing from about the 12th century BC. To consider the ancient mechanic who designed the Antikythera clockwork mechanism, performing on papyrus complex calculations of the size and number of required teeth on more than fifty wheels of the mechanism, strike us dumb with astonishment. Moreover he actually manufactured them, thereby carrying out the previous thousands of years of celestial observations to a precision of one ten thousandth. To carry out the reconstruction of the mechanism was hard task even in our laser and computer technology era.

From the beginning of the exploration of the surrounding physical world, in addition to the physical motion of objects originating in their mutual mechanical action (which men could see with their own eyes) man also meets with the existence of the phenomena of electricity and magnetism (which are to his own eyes invisible), understanding that fields operate by motion on the physical objects in their vicinity. In the 6th century BC there are written records of ancient Greeks that mention the magnetic properties of loadstone and electrical properties of amber (called an electron).

The first philosopher in the Greek tradition, the Greek sage Thales of Miletus (c. 624 – c. 546 BC), was historically recognized as the first individual in Western civilization known to have entertained and engaged in scientific philosophy. He set forth that nature is made of one substance apeiron (Greek word for ether) and that loadstone attracts iron because it has a soul.

In Greek mythology, ether represented a pure substance that the gods breathed. Aristotle claimed that only the natural motion of the ether as the fifth element (quintessence), which is located in the area of the sky, is its circulation in the circle and therefore stars also circulate along their celestial orbits--the likeliest correct explanation for the observed compact rotation of galaxies in the 20th century.

The new power-political structure after its accession in the first centuries of the new era claimed the Ptolemy (Claudius Ptolemy 90-168 AD, outstanding scholar of Alexandria) geocentric image of the universe as the official and the only tolerable image of the universe. The Ptolemy geocentric image of the universe was the sequel of the Greek philosopher Aristotle's (384 – 322 BC) geocentric universe from the 4th Century B.C., in which the fixed, spherical Earth is at the center, surrounded by concentric celestial spheres of planets and stars.

In 1600, Giordano Bruno was burned to death when, in lectures at Oxford, he claimed that the stars are remote Suns also surrounded by exoplanets and that the universe is infinite. The condemnation of Galileo in 1633 to life imprisonment was only thanks to the fact that the then Pope was a former friend of Galileo from their youth. This image of the Universe persisted until several decades after Kepler (1609 Astronomia nova).

The reason for the provisions of the Ptolemy geocentric image of universe as the official and the only tolerable image of the universe (though there existed the heliocentric view of the ancient Greeks), was its consistency with the idea of the principles of creation. The main proof of the correctness of the Ptolemy geocentric image was found in the mathematics (the complex geometry of the cycloids) of the calculation of the motion of other planets around the Earth. The complex and complicated math of cycloids, for which no one could give reasons from the physical point of view, was declared the finding of the order of the universe and the confirmation of the accuracy of the physical condition that the Earth was the center of the universe.

By this order of the Universe, celestial bodies in the Universe move (around the Earth) **without any material causation and their movement is a consequence of God's will, commanded by God's geometry of the mathematical structure.**

Ptolemy's mathematical description of the motion of the planets, even though based on an incorrect physical assumption, calculated the position of the planets on their orbits with even better precision than the nearly physically correct Copernicus (De Revolutionibus 1533) heliocentric model of planetary motion along circular orbits around the Sun. This was one of the reasons the Copernicus model was not accepted.

Kepler (Astronomia Nova 1609) after ten years of hard work finding the mystical mathematical formulas to explain the data of the movement of the planets from Tycho Brahe's precise astronomical observation, finally came to the simple mathematical rule of three that describes the movement of the planets along their elliptic orbits around the Sun.

Based on the heliocentric model, Kepler described the kinematics of movement of the planets by a simple mathematics and with better accuracy than Ptolemy. But he had not discovered the physical causes, that physical order by which the movement of the planets is governed in the solar system.

Kepler's simplification of the mathematical description of the motion of the planets to their mathematically trivial relationships allowed Newton to discover the physical cause and order, which is governed by the mathematics of kinematic description of the orbital system of the solar system. Newton discovered that the physical cause determining the order of movement of celestial bodies is the existence of a gravitational field as a real physical substance, existing in the surroundings of each ponderable body. This gravitational field around the ponderable body is inseparable from any ponderable body and Newton came to the general validity of the law of mutual gravitational interaction of all matter.

Newton explains the circulation of the planets around the Sun so that any two celestial bodies, through their own gravitational fields, attract each other (just as a falling apple from the tree also gravitates the Earth, although by negligible power). The forces of gravitational fields are well-balanced for a stable system of celestial constellations by centrifugal inertial forces on their mutual orbits and thus always circulate around a common center of gravity.

In the case of negligible mass of the planet to the mass of the Sun, this center of gravity is located inside the Sun, as a result of which the movement seems as if only the planet circulates around the Sun along the elliptical orbit. The planet actually orbits around the Sun, though not exactly around its center.

Newton his notion 'attraction' explained as - "*I here use the word attraction for any endeavor whatever made by bodies to approach each other; whether that endeavor arise from the action of the bodies themselves as tending mutually to, or agitating each other by spirits emitted; or whether it arises from the action of the ether or of the air or of any medium whatever, to enable acceleration or rotation to be looked upon as something real*" [1, p. 238].

Newton's 'attraction' is today considered rather as push forces of gradient of fields of the ether.

Newton attributed the gravitational forces, without any doubt, as so many of his contemporaries, to the existence of a force field as a real physical substance in the surrounding of each body.

"Gravity so that one body may act upon another at a distance thro' a Vacuum, without the Mediation of anything else, by and through which their Action and Force may be conveyed from one to another, is to me so great an Absurdity that I believe no Man who has in philosophical Matters a competent Faculty of thinking can ever fall into it. Gravity must be caused by an Agent acting constantly and according to certain laws" [3].

Evidently Newton was so strong a believer in the medium that we call the ether, that he was ready to discount the intelligence of any man who doubted it, though he could not work out its mode of action. Newton claimed that ether's adapted aethereal spirits produce the phenomena of electricity, magnetism, and gravitation.

This Newton belief is formulated in his main work, where it states that gravity is *"as a certain power or energy diffused from the center to all places around to move the bodies that are in them"*[1, p. 76]. Or also, as formulated in Newton's Letter to Robert Boyle [4] in 1678-9, *"I suppose, that there is diffused through all places an etherial substance, capable of contraction and dilatation, strongly elastic, and, in a word, much like air in all respects, but far more subtle"*.

This Newton's concept of subtle we can quantify, if we take into account his physical procedure for calculating the velocity of sound in the air. Then first estimation is a ratio of sound velocity in the air and light velocity in a vacuum so 10^{-6} of an air molecule. Second more likely estimation is their quadratic ratio 10^{-12} that corresponds to the mass of a neutrino. So the proton consist at least 10^5 particles of ether or more likely 10^{11} neutrinos.

R. Descartes (1596-1650), the father of modern western philosophy who had the most influence on Newton, considered space to be entirely filled with matter. **The formation of visual matter and planets, by Descartes, occurs from vortexes of ether. Descartes' vacuum of space is not empty but composed of huge swirling whirlpools of**

ethereal or fine matter, producing what would later be called gravitational effects [46].

In 1708 Newton wrote thus: *"Perhaps the whole frame of nature may be nothing but various contextures of some certain ethereal spirits or vapors, condensed, as it were, by precipitation; and after condensation wrought into various forms, at first by the immediate hand of the Creator, and ever after by the power of nature."*

Newton instead of this supposition and guess could have talked about surety if he had known that Maxwell's electrodynamics has brought us knowledge of the generation of electromagnetic waves of ether and that particle physics has brought us the knowledge of creation of solid mass particles from these electromagnetic waves (as is substantiated later in this book).

Newton's laws (gravity law as a gradient of force of medium, force law as resistance against acceleration in a medium, law of action and reaction, law of resistance of a body moving in a fluid proportional to the square of the speed of movement, the calculation of the speed of sound in the air and an estimate of the size of the elementary particles of air) were, for Newton, the particular steps in his effort to confirm the existence of this ethereal substance as it will be also referred later.

Newton already knew of the existence of the invisible phenomena of electricity and magnetism which, through force fields, causes motion among bodies and adds the phenomenon of gravitational forces. This is a much weaker phenomenon compared to electricity and magnetism in regards to the bigness of their source. At the same time, with the discovery of the gravitational field around any mass body, Newton (together with Galileo and other physicists) discovered the existence of the inertial forces that were also inseparable from any mass body.

Contemporary Physics, a hundred years since the inception of relativity, publicly repeats countless times the false claims about Newton's notion of mutual gravitational forces as the force between two mass bodies acting immediately and remotely through the void space of a vacuum with infinite speed. Newton, however, holds gravitational forces as the power through the medium and apparently assumes the final speed of gravitational forces in this medium, which may be the reason Newton interested himself and calculated the speed of sound in the air.

General relativity (1915), on the basis of forgery, claims Newton's idea of gravitational force as the forces acting at a distance through a vacuum and also conceals Riemann's necessary condition that curvature of space unavoidably requires material content filling it. GTR then finally brings to our civilization the allegedly greatest achievement of all history of the human spirit in understanding gravity as the curvature of non-material notions of space and time.

But as is detailed later in chapter 3, space time conception was fully refused by Einstein himself after 1920 -"*The theory of space (geometry) and time no longer represent intrinsic physics propounded independently of mechanics and gravitation* ".

The mathematics of Riemann's differential geometry of curved space became the basic language of the GTR, but Riemann himself, although a mathematician, asserts on the contrary (in 1854) that "*space in itself is nothing more than a three-dimensional manifold devoid of all form ; it acquires a definite form only through the advent of the material content filling it and determining its metric relations*" [5, p. 98]. Particles of matter which Riemann called Quanta represent, according to Riemann discrete portions, little hills, on a surface of a flat continuous manifold of space filled with matter.

GTR is based on the same concept of the description of gravity as a description of electromagnetism with a perspective of their integration. Thus the phenomena of electromagnetism (the electric field, magnetic field, electromagnetic fields) in parallel to GTR would also be assigned to the curvature of space and time. For nearly thirty years, from 1926 until his death in 1955, these were the central focus of Einstein's research, but his unified theory was an unmitigated disaster. No physicist was willing to admit that the electric, magnetic or electromagnetic fields are a curvature of non- material space and time.

Most of the physicists involved from 1760 to 1920 in the intensive exploration of electric and magnetic phenomena (Coulomb, Volta, Ampere, Orsted, Faraday, Ohm, Maxwell, Hertz, Edison, Weber, Tesla) fully believed in the existence of the electric and magnetic force fields as a real physical substance, called ether and they confirmed this substance in their experiments. This substance can spread waves of this substance caused by oscillations of the sources of gradients of fields of this substance. These waves can even spread independently of these sources and transmit inserted power (energy) into this substance.

Let us recall for all these physicists the statement of Maxwell who, in very last clause of his Treaties [6] (1873), declared: *"In fact, whenever energy is transmitted from one body to another, there must be a medium or substance in which the energy exists. . . . all theories lead to the conception of a medium in which that propagation takes place. . . and this has been my constant aim in this treatise".*

"Whatever problems we face in our attempts to work out a well-grounded notion of the ether structure, it is beyond any doubt that interplanetary and interstellar spaces are no empty spaces, they are filled with material substance or a body that is most extensive and, perhaps, most homogenous of those that are known to us".

In the modern era of physics from Descartes until the year 1905 there was not a single physicist who would have doubted an existence of the ether. All the most significant physicists of this period wrote articles or extensive works about the interaction of the matter with ether. This fact is well documented in book K.F. Schaffner, Nineteenth-Century Aether Theories [64].

The conviction of physicists at the end of the 19th-century of the full existence of ether can be also well seen in the research of H.A. Lorentz

- Ether theories and ether models (1901-1902), examining the work of many distinguished physicists of the 19th-century on ether (Stokes, Planck, Fresnel, Maxwell, Kelvin, Neumann). At the beginning of the 20th century the properties of the ether was a prominent subject of dissertations for doctoral degrees at most Universities in Europe. The introductions to most physical textbooks from the beginning of the 20th century include references to the existence of ether being a certainty.

Einstein, after 1916, came in with his rediscovery of the ether and he subsequently became a more enthusiastic advocate of the proven existence of the ether than supporters of the ether before 1905. In his papers (e.g. 1916, 1920, 1924 discussed more below) Einstein claimed: *"According to the general theory of relativity space without ether is unthinkable. The mechanical ether, designated by Newton as 'absolute space', must therefore be considered by us as a physical reality. In Newton's theory of motion, space has physical reality - in contrast to the case of geometry and kinematics. ....ether has to serve as medium for the effects of inertia ... now it appears that ether will have to be regarded as a primary thing and that matter is derived from it... We are not going to be able to dispense with the ether in theoretical physics. The theory of space (geometry) and time no longer represent intrinsic physics propounded independently of mechanics and gravitation. According to our present conceptions the elementary particles of matter are also, in their essence, nothing else than condensations of the electromagnetic field".*

But this, Einstein's final cognition, is concealed from us and gravity as the mystery of nonmaterial space time curvature is daily forced upon the public and the wider physical community.

In 1925, Edwin Hubble announced his evidence confirming that the bright fog formations in the night sky (in the meantime called nebulas) are separate groupings of stars and galaxies and that all the other stars we observe in the night sky, free and with our own eyes, belong to our Galaxy, the Milky Way.

For proof of the theory of the Big Bang, current physics considers increasing the red shift with the distance of galaxies, measured in the spectra of galaxies first by Hubble (1927-29). Hubble himself, even when he was pressured (mainly by Lemaitre at the IAU meeting, 1928), disapproved of this unilateral interpretation until the end of his life. The Nobel Prize for astronomy till the 1950s was not granted, and so Hubble did not have to succumb to this pressure (Unlike Millikan in 1921).

Hubble, for a more likely explanation than explaining the red shift spectra by mutual receding of galaxies, considered the explanation of this shift by the loss of light energy passing through the medium of interstellar space. We can cite from the work of Hubble [7, p. 1] (1937), The Observational Approach to Cosmology,

"The features, however, include the phenomena of red-shifts whose significance is still uncertain. Alternative interpretations are possible, and, while they introduce only minor differences in the picture of the observable region, they lead to totally different conceptions of the universe itself".

„The cautious observer naturally examines other possibilities before accepting the proposition, even as a working hypothesis. He (Hubble) *recalls the alternative formulation of the law of red-shifts - light loses energy in proportion to the distance it travels through space. The law, in this form, sounds quite plausible. Interior nebular space, we believe, cannot be entirely empty".*

We can also cite from work A. K. T. Assis et all [8] - Hubble's Cosmology: From a Finite Expanding Universe to a Static Endless Universe (2011) -

"We show, by quoting his works, that Hubble remained cautiously against the big bang until the end of his life, contrary to the statements of many modern authors".

The consequence of this paper should have resulted at immediate removal or at least suspension of the Big Bang theory in physics.

Even today, the hundreds of non-fiction documentary films of the most respected television or most respected Web sources of information dedicated to the description of the evolution of opinion of mankind on the physical image of the universe state that Hubble's observations are evidence of the expanding universe. See Wikipedia:
- Big bang https://en.wikipedia.org/wiki/Big_Bang "*In 1929, from analysis of galactic redshifts, Edwin Hubble concluded that galaxies are drifting apart, important observational evidence consistent with the hypothesis of an expanding universe*".
- Edwin Hubble https://en.wikipedia.org/wiki/Edwin_Hubble "*Hubble is known for showing that the recessional velocity of a galaxy increases with its distance from the earth, implying the universe is expanding*".
The Belgian priest Lemaitre proposed in 1927 the theory of the expansion of the universe, widely misattributed to Edwin Hubble. He was the first to derive what is now known as Hubble's law v=HD and proposed H what is now called the Hubble constant. But the relation v=HD is taught at schools all over the world as Hubble's law, with H as Hubble's constant.

Added to that, Hubble himself explains red shift by the loss of light energy passing through the medium of interstellar space. This is not mentioned and is concealed to students.

In 1913 the structure of matter into atoms (Perin) and in 1920 into protons (Rutherford) was experimentally confirmed.

Physics brought full proof of the existence of the ether, in 1932 (Anderson) and in 1955 (Laboratories in Berkeley), with the discovery of the production of pairs of particles and antiparticles of electrons and protons from electromagnetic (etherial) radiation.

With the mechanical waving of a magnet nearby copper wire we produce electromagnetic radiation with a frequency equal to the frequency of the waving magnet. At sufficient frequencies, the mass of pairs of electrons or protons can be produced from electromagnetic radiation. For an arbitrarily long time we can do this waving with a magnet and produce any amount of photons, electrons or protons, but from the magnet or wire wane not even a piece of mass.

This physical fact can in no way be explained by force fields theories using force-mediating particles.

On the contrary, at the annihilation of these particles with antiparticles, two photons of electromagnetic radiation arise. These two photons in subsequent scattering on atoms, e.g. steel ball in void space, transfer its energy to this ball and completely dissipate into nothingness. This steel ball, after a short warm up from the photons, cools down again at the temperature of the universe around -270° C. This means we have under current physics, right before our eyes, an experiment concerning the invalidity of the law of conservation of mass and energy.

The body of an astronaut, if leaving the craft into the void space of the universe without a space suit, would immediately freeze to a temperature close to absolute zero. His or her thermal energy disappears although according to today's physics, no air or material substance is situated nearby. This loss of thermal energy must be related to thermal radiation, as all matter with a temperature greater than absolute zero emits thermal electromagnetic radiation.

But what substance are astronaut bodies balanced to in thermodynamic equilibrium (balance between two ambients) at a temperature around -270° C? No doubt this temperature is different in separate areas of space so the reverse process must also exist, when bodies translocate from one temperature to another.

We have created the energy of the electromagnetic field, and subsequent mass of the electron or the proton under current physics from nothingness--waving a magnet around a wire. This matter by annihilation into electromagnetic radiation energy afterwards disappeared before our eyes into nothingness by scattering.

The creation of pairs of particles and antiparticles is not a limited phenomenon of physicists in laboratories, but is a common and well examined phenomenon of the interaction of electromagnetic radiation with matter. Electromagnetic radiation up to an energy of 1.02 MeV interacts with matter in photoelectric effect or scattering processes (Compton, Rayleigh scattering). For interactions over 1.02 MeV and up to 1.9 GeV, electron positron pair creation predominates and over 1.9 GeV outweighs the creation of proton antiproton pairs.

In the universe and nature all around us on Earth this phenomenon is continuously going on in a great quantity from the gamma radiation of radionuclides present to a greater or lesser extent in every substance on the ground (up to 20 MeV), as well as from storm lightning (100 MeV) and the high energy gamma radiation (80 GeV to millions TeV) incidents on our Earth in great quantity from the universe.

So, the annihilation of protons and antiprotons was confirmed experimentally, which is the conversion of mass into an electromagnetic curl of the ether and vice versa.

But current physics claims in the Higgs field theory that the Higgs boson is the *only* method by which all particles of matter in the universe acquired their mass.

Perhaps current physics does not want us to claim that, in creation of the proton from the electromagnetic radiation at energy 1.9 GeV or vice versa in the process of conversion of a proton into the 0.94 GeV electromagnetic radiation between the proton (0.94 GeV) and electromagnetic radiation stands energy more than a hundred times greater 125 GeV Higgs boson.

Perhaps current physics in the Standard model does not want us to claim that, before process of creation of proton - antiproton pairs (or all other particles-antiparticles pairs) from electromagnetic waves, some of the three free quarks or of three free antiquarks (later synthetized) exist in photons of these electromagnetic waves. Or perhaps it also does not want us to claim that, after the annihilation of proton-antiproton pairs, these quarks are separated and somewhere exist or vanish into two photon of electromagnetic waves. But for separation, so also for the synthesis of quarks, an infinite amount of energy is necessary according to the Standard model.

In 1964, Gell-Mann introduced the purported existence of quarks as particles of which the hadrons, as parts of an ordering scheme for hadrons, are composed, even if **there *was no evidence* for their physical existence**. Gell-Mann conceived of a mysterious physically inconsistent principle that quarks can never be directly observed or found in isolation, because an infinitely huge amount of power is necessary for their separation. This unproven speculation includes, in itself, the impossibility to uproot it.

Later in 1968 it was declared that accelerator experiments of inelastic electron-proton scattering at Stanford Linear Accelerator Center (SLAC) **allegedly provide evidence for the existence of quarks**. The main work referred to for this allegedly provided evidence for the existence of quarks is the outstanding researcher at Stanford Linear Accelerator Center, J. D. Bjorken. However, Bjorken in 1969 declared on page 4 in his paper [9] that *"There are various theoretical models which try to explain or at least describe these features of data but none work really well, or are totally satisfying. We will discuss three of these theoretical descriptions of the data; these are: 1) incoherent scattering from pointlike constituents within the proton – the parton model, or Thomson nucleon, 2) vector dominance, or Rutherford electron, 3) current commutators"*.

In Bjorken's paper, no clear advantage for any model is provided. Last but not least, it should be noted that in all models the electron is taken as an approximately dimensionless point-like probe, which is opposed in our previous paper [10, 11].

At SLAC experiments, the only evidence for the consideration of the alleged existence of quarks in the interior of protons is the detected asymmetry of the field around protons in these inelastic, electrons impinging upon protons, scattering experiments. But it is necessary to mention the fact that electromagnetic fields between atoms and molecules of fluids (liquid of the hydrogen was the target at SLAC experiment) configure themselves in polygonal patterns. Each atom is surrounded by at least another three atoms, which most likely also results in partial asymmetrical non-spherical shape of protons itself.

This resulted in a different amount of scattered electrons around protons in the different directions measured at SLAC. We can claim that the asymmetry of fields identified in the vicinity of protons of liquid hydrogen at SLAC inelastic electron-proton scattering experiments was not a consequence of the internal structure of protons but a consequence of the asymmetrical patterns of the external electromagnetic fields in the closest vicinity around protons of the hydrogen liquid.

In 1970s, masses of Gell-Mann's three quarks as the scattering centers, from which were allegedly formed a proton, corresponded roughly to the one third (around 313MeV) of a proton mass 940MeV. They represented spherical ball with a diameter around one third of proton and did not have any further dynamical meaning. Today it is not the true and this idea was thrown out of physics. These Gell-Mann's quarks sank into history and are called now as 'constituent quark masses'. Under the pressure of the next experimental data significantly different quarks from Gell-Mann's quarks were introduced and allegedly proved by the mathematical machinery of re-normalization. Correct quarks masses which allegedly today constitute the proton, called as 'current quark masses', are just a one hundredth (in average 3,6MeV and with a diameter around one hundred of proton's) of original Gell-Mann's quarks and so represent just one hundredth of the proton mass. The model marked in the above-mentioned Bjorken's paper as 1) - the parton model (partons fall in with quarks) – so Gell-Mann's quark model awarded with Nobel Prize was tacitly abandoned and model marked in Bjorken's paper as 3) - current commutators model - is used today. These hypothetical today's quarks (never observed or measured) have no physical shape, they represent whirling points which current or flow together with a hypothetical (never observed or measured) gluons in so-called quark-gluon plasma all over the room of the proton. Nothing would impede the eventual today's assertion that there are not three but six or nine or other quantities of quarks in three or more currents in proton. As current Standard Model considers a dimension of one quark as the one hundredth of proton's dimension, then up to million quarks can be present in one proton. This is consistent with our estimation stated above.

Just a fairy tale about the discovery of three quarks, Gell-Mann's Nobel Prize award and ordinance of power structure keep the number of quarks in proton on three. This way Gell-Mann's quarks vanished out of a proton and just the word quark remained. These today's quarks have no experimental, physical or logical justification. The purpose of their abidance in theories is just in that something must be a bearer of mass because allegedly gluons are mass-less in analogy to the nonsense installation of mass-less photons in relativity and allegedly the ether does not exist.

The main role in Standard Model fairy tales developed in 1970s plays never observed or experimentally measured hypothetical gluon (the mediator of strong forces as an exchange particle between hypothetical quarks) which exists allegedly in eight forms of different color. But as in last two decades quarks mass was one hundred fold reduced and reverted into joint currents with gluons all fiction of the Gell-Mann's quark model, awarded with Nobel Prize about gluons as an intermediate particle between static quarks (scattering centers at SLAC 1968 scattering experiments) break up. Among others, also the explanation of force fields in the 20th century on the basis of exchange of particles between two objects is fully absurd physics.

Finally, inside produced quark-gluon plasma in 2015 CERN experiment [15], detailed below, none of these new current quarks or any other particles which were supposed to exist there were observed and so quarks vanished out of physics totally. Just the fairy tale of power structures about the alleged existence of Gell-Mann's inseparable holy trinity quarks inside a proton, forced by mass media on public, remained. The evanishment of quarks inseparably means also the evanishment of the whole Standard Model together with the Higgs boson.

The robust fantastic theory of the so called Standard Model enabled mysterious physical properties (as fractions of unit electrical charge and their different ratios to mass, infinitely huge power for separation) was generated in the mid-1970s to accommodate the results. Later and whenever necessary, go-as-you-please other mysterious physical properties were fabricated into this model.

T. Ferbel in his text book [12] states that the Standard Model has many parameters, e.g., masses of the leptons, quarks, gauge bosons, and of the Higgs, various coupling strengths and elements of the CKM matrix, with all values seemingly perplexing and ad hoc.

Ferbel, in his presentation [13] in 2012- Belief and Observation: The Top Quark and Other Tales of 'Discovery', describes his personal adventures with keen physicists at SLAC experiments searching for top quarks who verbalized their approach to experimental work as, *"I'll find top, even if it's not there!"*

The so called Standard Model contains quite a large number of theories [14]. Physicists complain about these theories because the simplest one has 19 adjustable constants and the more elaborate version has 29 adjustable constants. But constants in physics represent a calibration point of physical law, so each of these 19 up to 29 constants represents unknown physics. Yet the physicist claims that the Standard Model is the best theory in the science of particle physics.

Physicists at CERN announced to world in 2012 that their experimental results of one hazy hump (increase amount of events) on smooth curve through these 29 adjustable constants without doubt clearly points at their single one primordially picked physical model from tens of others and that so Higgs boson was discovered and that thus even the existence of Higgs boson was confirmed.

This detected one hump through these 29 adjustable constants can point at to at least another 29 primordially picked physical models explaining the measured data. On top of that the mass of Higgs boson, which the physicist at CERN allegedly discovered, does not fit to any one of these theories based on 19 up to 29 adjustable constants contained in the Standard Model.

The claim that the standard Higgs boson model is a single correct model, without considering the correctness of other physical models, has nothing to do with the scientific methods in physics. It is pure tautology, obscuring the lack of evidence or valid reasoning supporting the stated conclusion. It also was not determined whether this is a resonance or particle by the following collision or decay experiments with this allegedly discovered particle, as is usual for confirming the discovery of a particle in particles physics. On top of that, physicists at CERN simply declared discovery of the Higgs boson and the Nobel Prize was immediately awarded for it, **but they do not know its lifetime! which is merely predicted to** $10^{-22}s$.

For more details we refer to the Alexander Unzicker 2014 book 'The Higgs Fake: How Particle Physicists Fooled the Nobel Committee' [56]. The main conclusions of his book is: *"The field theory cornerstones of modern particle physics are based on a completely metaphorical level, upon which you may develop any fantasy to explain your needs. The standard model is incapable of assigning a meaning to the mass of a particle, just as astrology has nothing to say about a star's luminosity. It is an arbitrary construction that will leave anybody frustrated who seeks insight.* ***Maybe one cannot understand Nature,***

but for sure, particle physics cannot explain it. *It is time to dump a big science enterprise that has grown to absurd complication in every sense, has swept under the rug the important problems, has developed nothing useful and impedes any true progress in understanding the laws of Nature. It is obvious that opinions in high energy physics are homogenized by social and hierarchical pressure. It is no excuse that, unfortunately, there are other degenerations of the scientific method in the realm of theoretical physics: supersymmetry, and string theory which never predicted anything about anything and never will.* **Therefore, it is time to stop seeing the Nobel Prize as a sacrosanct accolade for physics. In the course of the last 50 years, the award has contributed considerably to the degeneration of the search for the fundamental laws of Nature.** *"*

Symptomatic for approach of most 'genial' physicists to understanding the laws of Nature is the frequently cited announcement of Gell-Mann about theoretical models in particles physics: *"We all know how to use and how to apply it to problems; and so we have learned to live with the fact that nobody can understand it"*.

It is hearsay that the physicists at CERN in fact formed a division with the different opinion to CREN official opinions kept among themselves because they are afraid of losing their jobs. Concerning the veracity of the allegedly discovered Higgs boson from the 17 principle investigators at CERN, 15 of the 17 said that **they do *not* think they had found the Higgs boson** and two said they did.

Within the mainstream scientific community half of physicists judge that Higgs was not discovered and particles such as Higgs do not even exist. How is it possible that, **without any defense before the scientific community**, the discovery of the Higgs boson is simply declared and the Nobel Prize is immediately awarded for it? But for the CERN budget of 1billion euro per year (equals around to the Gross Domestic Product of Liberia with 4 million citizens) it is unthinkable not to return the breakthrough results, **no matter if they are true or not**.

In 1971 Kuti and Waisskopf, in a nucleon model in addition to the three quarks, requested a sea of quark-antiquark pairs and neutral gluons for the composition of a nucleon.

In the last decade and **based on experimental results**, physics came to the conclusion that **the mass of particles** (till then asserted as a static composition of quarks with gluon fields in elementary particles) **lies in the spinning quark-gluon field and the actual mass of the quarks makes but a minimal contribution to the mass of the particles**.

In 2015 another international team of physicists at CERN had produced quark-gluon plasma at the Large Hadron Collider by colliding protons with lead nuclei at high energy. **They contrarily reported [15] that this state of matter doesn't (as was initially expected) behave like a gas of quarks and gluons, but rather like a continuous liquid (no quarks or any particles).**

This report on the absence of quarks (which stand in the hierarchy of the Standard Model on the bottom level) means also the absence of the Higss boson (which stands in the hierarchy of the Standard Model on the supreme level) conjoined firmly to existence of quarks.

The consequence of this report [15] should have resulted in the immediate removal or, at least, suspension of the Standard Model.

The lifetime of all (around a hundred) so called elementary particles, except stable protons, electrons, photons or possibly neutrinos (or neutron max 15min) is one-millionth of a second to a billionth of a billionth of second (hyperons from $10^{-10}s$ to $10^{-20}s$, mesons from $10^{-8}s$ to $10^{-20}s$, leptons - muon $10^{-6}s$, tauon $10^{-13}s$).

All decays of hyperons from the largest energy through less energy terminate at protons (as do neutrons) and energy is washed away by neutrinos, photons or electrons.

All decays of mesons (composed allegedly of two quarks) from largest energy through less energy terminate (via leptons) at electrons (containing no quarks) and energy is washed away by neutrinos or photons, whereas quarks at these decays simply vanish without any physical reasoning of how and where, though infinite energy is allegedly necessary for their separation.

Lepton decays (muon and tauon) - terminate at electrons and energy is washed away by neutrinos.

Thus vice versa, we can say that all leptons and mesons are a series of excited energy states (more stable) or resonance (less stable) of electrons (or positrons). Hyperons are a series of excited energy states or resonance of protons (or antiprotons). We can say that this is the most natural and physically simple first approach to the primordial model and classification of so called elementary particles.

The example of the full absurdity of the Standard Model can be presented in a simple physical and logical situation when a pi meson, pion, simply decays (reduces its energy and becomes mu meson) to a muon and energy is washed away by a neutrino. No other differences between these two particles then less mass was ever measured. The muon was originally, according to experimenters, simply less heavy pi meson and was called originally mu meson. In order to implement quarks and break this simply conclusion from memories of physicists was renamed by theorists as muon.

This simple experimental case in fantasy of Standard Model looks another way.

According to the fairy tales of the Standard Model, at final lap of mesons decay, from the pi meson with energy 140MeV is allegedly created (without any external interference or contribution) W boson with robust energy 80 000Mev ! containing allegedly the two quarks from pi meson – up 2,4MeV and down 4,8MeV. W boson than allegedly decays into a muon with energy 106Mev and a neutrino with 0,1eV. From alleged two quarks in W boson the one original quark allegedly transmutes to the antipodes of the second and thereafter these two quarks scentless annihilate!

This scentless annihilation is in controversy with all experimental observations of annihilation of particles and antiparticles which products are always two electro-magnetic photons. It is clear for everyone that also trivial math of energy balances is false in this Standard Model pion decay reasoning.

No quarks exist. No Higgs boson exists, because there is no reason why other excited states or the resonances of proton energy series in higher energy ranges above energy detected in 2012 at CERN hereafter could not be found.

In conclusion, it is necessary to emphasize the most important, firm and generally known fact which disappears in the flood of the presumed, often trumped-up, 'elementary' particles and hypothetical building elements of the matter. All the rigid matter of galaxies, stars, planets, the animate and inanimate nature, all atoms, so all elements of the Mendeleev periodic table are formed from the proton or clusters of protons whose positive electric fields are screened by a cloud of (curled opposite force field of ether) shells called electrons. The neutron is also the proton, screened by an electron cloud, since free neutron decays into the proton, the electron and the antineutrino.

3. Einstein's Destruction of Physics and Scientific Principles

Einstein (1879 -1955) from 1887 till 1894 studied for 7 years at the Luitpold Gymnasium in Munich but he did not finish, convincing the school to let him go by using a doctor's note (Dr. Talmud) claiming nervous exhaustion [44].

In 1895, Einstein failed a simple entrance exam to Zürich Polytechnic, as he intended becoming a secondary school (Gymnasium) teacher. He attended gymnasium in Aarau, Switzerland, in 1895–96 to complete his secondary schooling. In January 1896 Einstein renounced his citizenship in the German Kingdom to avoid military service.

In 1896 Einstein enrolled in the four-year mathematics and physics teaching diploma program at the Federal Polytechnic School in Zürich. Zürich Polytechnic School was upgraded to a school of the university type renamed as Federal Technical High-school some 15 years later in 1911, with the right to grant graduate degrees. **After two years study at Zürich Polytechnic School, Einstein failed a basic physics course of Physical experiments for beginners, scheduled for students during first two years.**

Obtaining experience about real-world of experimental physics discomforted him. One of his professors called him a lazy dog [44]. In March, 1899 Einstein was given an official director's **reprimand due to lack of diligence in physics practicum**. His low-ranked 4.9 average mark **was just enough to let him get his diploma**. In 1900, he was awarded a teaching diploma.

By graduating the Polytechnic School Einstein retrieved insufficient education (e.g. Maxwell's theory was not covered in school's lectures [50]) and retrieved insufficient experimental experience for the work in theoretical physics that he was trying to make later in his life.

In 1900, at 21, Einstein obtained his first wrist watch. At that time (no radios and televisions as well as no other radio signal transmitters

41

existed, telephone was commonly utilized after 1910) the adjusting and synchronizing of clocks (simultaneity) was a weighty problem. From around 1880 in Europe the synchronizing of clocks was provided by transmitting time signals via telegraph lines to railway stations. For ordinary residents without contact to the railway station hearing passing by the train in regular time of a day was a most common method of clock synchronizing. No automobiles existed and the train was utilized for transport as the highest speed conveyor. That is why in his STR thought experiments Einstein always stands at the train station and fantasizes about synchronization of clocks by speed of light.

In 1911, in Prague, in his 32nd year, Einstein and his family had electric lighting in their home for the first time. Five per cent of Berlin's homes boasted electric power in 1914.

After obtaining his teaching diploma in 1900, Einstein spent almost two frustrating years searching for a teaching post. With the help of family friends he at last obtained his first job. From 1902 till 1909 he was a technical expert third class at the Swiss Patent Office, which meant that he was incompetent for a higher qualified position [26]. He and his own family (in 1903 he married M. Maric, they separated in 1914 and divorced in 1919) were in a permanently distressed financial situation. Einstein tried to change this situation by producing and publishing an excessive number of fantasies, he called theories, from behind his patent clerk desk. All his fantasy theories immediately aroused conflicting controversy from all the great physicists of that time.

Einstein intended to work for a doctoral degree at the University of Zurich under H. F. Weber on a topic related to thermoelectricity, but Weber refused him. The properties of the ether or the kinetic theory of gases were the prominent subjects of student dissertations. Einstein submitted a dissertation on molecular forces in gases to the University of Zurich in 1901, about a year after graduation from the Zürich Polytechnic, but withdrew it early in 1902 in order to avoid controversy with Boltzmann. Three years later in 1905, after Boltzmann left Germany to Austria, the dissertation was again submitted.

Boltzmann was the most significant physicist in these topics and published many works after he received his PhD degree in 1866 for his dissertation on the kinetic theory of gases.

In January, 1906, Einstein's thesis was accepted. On the 15th of January, 1906 he was awarded a doctoral degree and thus optically upgraded his Polytechnic School non-high school education to the high school level. The Annalen der Physik received a different version of his thesis for publication. Einstein **corrected this publication** from 1905 in 1906 in a published supplement to the thesis [12, V2, D15, D33].

In 1911 Jacques Bancelin performed experiments in Perrin's laboratory and found a significant **discrepancy** between the results of his experiments and Einstein's predictions in his 1905 published and later in 1906 already once corrected dissertation paper. **A calculational error in Einstein's 1906 paper was announced**. Einstein himself was not able to find the error in his calculations and, in 1911, asked his student and collaborator Ludwig Hopf to check the calculations. *"I rechecked my old calculations and arguments and could find no errors in them. You would do a great service to the cause if you made a thorough examination of my arguments"* [12, V8, D239].

Correction of the error, which was found by Hopf, is published in Einstein's 1911 *"Correction to My Paper: A New determination of molecular dimensions"* [12, V3, D14]. In introduction to this paper Einstein thanked Hopf for finding the errors. This was the **second correction paper of his 1905 paper**. This correction was reiterated in Einstein's 1920 paper and integrated into the republication of Einstein's dissertation in Einstein 1922 [12, V2, John Stachel, Einstein's dissertation on the determination of molecular dimensions].

In connection with relativity it is symptomatic that errors was found in derivatives of velocity components, which occur in equations for pressure components. This proves that Einstein at least until 1911 did not understand the physical nature of quantities of classical mechanics extracted from physical experiments - that the change of velocities, the derivatives of velocities, result in accelerations, so in the forces so in change of pressure ratios, so in changes of densities. That is why he did not use these quantities of classical mechanics in his reasoning in the construction of his 'new physical laws' (1905 STR) at least until 1911. This way he constructed his new physical laws of his fantasy land based on linear addition and subtraction of velocities at Galileo's transformation which are in contradiction with experimental laws of classical mechanics based on accelerations, forces and kinetic energies.

This situation persisted even in 1913 (detailed below) when M. Besso, who was a coworker on Einstein's "Entwurf equations" of gravity, taught Einstein physical reality and showed him that his approach based on Machian idea of gravitational potential as a rotation of the fixed stars is physically wrong due to revealing of Coriolis forces at decreasing diameter of rotation of a 'hollow' or due to increasing inertial force with increasing acceleration in the void space [65].

By linking at will the Lorentz factor to classic physical laws at different reference frames, Einstein in STR produced a number of 'new physical laws' for reliance of reference frames (moving bodies) to their velocity. But there was never solved or even raised in STR the question of how bodies, inertial frames, receive their velocities, which was the central question of classical physics from which laws of classical physics was derived.

Such a controversial procedure as with his dissertation was inextricably linked with the whole scientific, theoretical, fantasy work of Einstein's career from its beginning till the end.

In 1907 Einstein published a paper 'On the relativity principle and the conclusion drawn from it'. One year later in 1908 he published paper titled - **Correction to the paper: "On the relativity principle and the conclusion drawn from it"** on which 2 pages around 10 relations from the 1907 paper were corrected [12, V2, D47, D49].

In January 1907 Einstein published a paper on Planck's theory of radiation and the theory of specific heat and after Planck's objection, in March 1907, Einstein published a **correction paper** "Correction to my paper: Planck's theory of radiation, etc" [12, V2, D38, D42].

Such a controversial procedure was also the case in his April 1908 paper 'On the fundamental electromagnetic equations for moving bodies' which was half a year later corrected in 3 relations in the paper of August 1908 - **Correction to the paper: "On the fundamental electromagnetic equations for moving bodies"**. After Max Laue (Nobel prize in 1914) showed mistakes in Einstein's last paper, Einstein in a **second correction paper** published in December 1908 [12, V2, D51, D53, D54] corrected the previous corrected relations in around 10 relations - Remarks on our paper "On the fundamental electromagnetic equations for moving bodies".

Such a controversial procedure was also the case of Einstein's 1915 paper "Experimental Proof of Ampere's Molecular Currents". This was corrected in the same year in his paper "Correction of my joint paper with J.W. de Haas: Experimental Proof of Ampere's Molecular Currents" [12, V6, D15, D16].

Such a controversial procedure was also the case of Einstein's 1905 paper [12, V2, D23], the so called central work of special relativity 'On the electrodynamics of moving bodies'.

This Einstein paper contains not a single reference although until 1905 there existed far more than 10 papers with similar or equal contents and even with almost the same title to Einstein's 1905 paper, written by a physicist who really deeply understood these topics, unlike Einstein. From them we can mention mainly-

Thompson 1881, On the Electric and Magnetic Effects produced by the Motion of Electrified Bodies

Voigt 1887, On the Principle of Doppler

Heaviside 1889, Electromagnetic waves, the propagation of potential, and the electromagnetic effects of a moving charge

Lorentz 1895, Attempt of a Theory of Electrical and Optical Phenomena in Moving Bodies

Poincaré 1898, The Measure of Time

Lorentz 1899, Simplified Theory of Electrical and Optical Phenomena in Moving Systems

Cohn 1901, On the equations of the electromagnetic field for moving bodies

Wien 1904, On the differential equations of the electrodynamics for moving bodies

Cohn 1904, On the Electrodynamics of Moving Systems

-and many others.

The fact that the reason and result of Einstein's 1905 paper was to bring a solution in the **mystery of nonphysical time manipulation, which replaced the Lorentz-FitzGerald contraction**, is confirmed in Einstein's 60 page 1907 paper 'On the relativity principle and the conclusion drawn from it' [12, V2, D47]. In the paper we can read *"Michelson-Morley (M-M) experiments contradiction was removed by Lorentz and FitzGerald ad hoc postulate of a **certain** contraction of moving bodies as artificial means of saving the theory.*

Surprisingly it turned out that a sufficiently sharpened new conception of time was all that was needed to overcome the contradiction*...the conception of a luminiferous ether does not fit into*

*this conception ..**therefore Lorentz's - FitzGerald theory should be abandoned…***".

Lorentz's – FitzGerald's clear physical reasoning at M-M experiment, consisting in the conclusion that dimensions of solid bodies are slightly altered under the pressure of the ether by their motion through the ether, Einstein considered as an artificial means, a kind of ad hoc postulate and contrary to his non-physical (time is not a physical notion and is a subsidiary notion to movement) **metaphysical explanation in deformation of notion of time he self-praised as non-artificial**. How ridiculous!

The purpose of Einstein's dilation of time was to bring an adverse solution to the length contraction solution of Lorentz and FitzGerald. Purported null results of M-M experiments using the Lorentz factor can be explained **by either the contraction of length or the time dilation, but not by both at the same time**.

But Einstein declared later at least from a 1911 paper [12, V3, D238] or e.g. explicitly in his 1920 book Relativity [35] " *…by means of Lorentz transformation…the rigid rod is shorter in motion..* **If we had based our considerations on the Galilei transformation we should not have obtained a contraction of the rod as a consequence of its motion**".

Einstein here controversially renounced his first physical approach in STR from 1905 -1907 based on the Galilei transformation which was Einstein's basic approach to derive STR laws.

So later Einstein, as well as today's official physics brings the opposite claim to his conclusion from 1905-1907 that Lorentz - FitzGerald theory should be abandoned. According to Einstein after 1911 both the length contraction of rigid bodies jointly with time dilatation are real physical results of STR although purported null results of M-M experiments cannot be explained by both. So, although

Einstein from the position of his ingenious 'sharpened new conception of time', based on the Galilei transformation in 1907, disapproved Lorentz and FitzGerald's conception as 'artificial means of a certain contraction of moving bodies', this Lorentz artificial means became a dominant conception of his later 'improved' theory of relativity.

In Einstein's 1905 paper, lengths of measuring rigid rods undergo no changes in different inertial frames –*"In accordance with the principle of relativity, the length of the rod in the moving system must be equal to the length l of the stationary rod"*. The so called length contraction in this paper subsists in different measurement results of lengths at different velocities of frames. Relativity in the 1905 paper lays in different measurement results arising from measurements of the time, simultaneity, length and velocities in inertial frames at different uniform translation velocities. As the different results in measuring of lengths flow from the time simultaneity and the different results in measuring of velocities flow from different results in measuring of run of times, in the 1905 paper (as is confirmed in the 1907 paper) all results end in measuring different results of times.

Till this point, the requirements of STR that length of bodies, length of space intervals, time intervals remain in all inertial frames the same appear as natural and **STR transformation relations simply provide the same results in different reference frames**.

But the absurdity of STR arises from the postulate that in all reference frames, regardless of their velocities, the measurement of velocity of light must provide the same results of the ultimate and always constant velocity of the speed of light (this postulate of light constant velocity Einstein recalled in 1913, as is detailed below). This physical absurdity is in STR afterwards attained in Einstein's 1907 paper [12, V2, D47] by the controversial stunt claim that transformation of time intervals are no longer different measurement results in different frames, **but are their own physical reality**. Frames with different velocity have a different clock rate and the notion of simultaneity in the direction of motion is altered, which results in the absurd addition law for velocities. Afterwards space neither exists independently of a physical reference frame nor is associated with a privileged reference frame.

In the 1905 paper the propagation of light in empty space had standard (so as waves in the water or in the air) natural requirements of physics –"*The light is always propagated in empty space with a definite velocity c which is independent of the state of motion of the emitting body*". But Einstein, in his absurd addition law for velocities, in which by change of velocities also time is changed which results in $1+1 \neq 2$, come to absurdity that mutual velocities of all moving bodies or photons is always c (more detailed in chapter 4.).

The separation of quantities of the velocity and time as independent quantities is the greatest disaster and fraud of STR and GTR. In physics and, primarily, in mechanics the time and movement and the time unit and the unit velocity are the same notion and cannot be separated, as is explained in chapter 5. of this book.

As is shown below in this chapter all these STR claims from 1905 resp. 1907 were recalled by Einstein from 1911 till 1934.

Another situation in the row of simultaneous validity of two opposite claims in STR is the claim that mass in reliance with velocity is increasing and the simultaneous claim that rest mass is an invariant in all reference frames and does not increase with the increasing velocity of frames.

In Einstein's 60 page 1907 paper [12, V2, D47], bodies in different inertial frames are rigid without any changes and all Einstein's physics consists in changes of his new time, time coordinates, clocks, which results in his new absurd addition of velocities. Mass of bodies as well as electric mass is stated as independent of the state of motion of the reference frame and are constants at any reference frame. In this paper, change of frequencies in optics are a consequence of Einstein's **newly introduced and calculated group velocities of bodies**.

But later in his papers yet in 1911 [12, V3, D23] Einstein's basic argument is - "*inertial mass of a body falling in gravity is increasing and this must be equal to increase in its gravity mass, otherwise bodies would not fall with the same acceleration and Galileo's law would be not valid*".

Subsequently, it is claimed that acceleration at fall of bodies in gravity can by simulated by acceleration of these bodies in void space. So increasing of inertial mass at acceleration generally is firm first fraction of this claim contrary to claims in STR from 1905-1907.

But this last affirmation Einstein sequentially fully uproots in his perhaps best known thought experiments presented in 1913 paper [12, V4, D13] right in the first clause – "*An observer enclosed in box can in no way decide whether the box is at rest in a static gravitational field, or whether it is in accelerated motion, maintained by forces acting on the box, in a space that is free of gravitational fields (equivalence hypothesis)*".

This contradicts the 1911 paper that the 'weight' (the pressure at the bottom of the box) of the accelerated observer enclosed in box will increase while his weight in the static gravitational field remains constant. So there is no problem for an observer enclosed in Einstein's box to decide whether the box is at rest in a static gravity field or whether it is in accelerated motion, maintained by forces acting on the box. Simply sensitive enough scales are needed. In this book, in section 4. we bring a closer view of this schizoid Einstein logic.

Here we clearly see that Einstein claimed two mutually exclusive contrary claims that prevail for him across the whole of his relativity theory.

Einstein's reasoning at introducing basic axioms at STR is contradictory from a physical point of view and impossible from the logical point of view.

In his 1907 paper he states -*"What is the influence of (uniform) acceleration on the shape of the body? **If such an influence is present, it will consist of a constant-ratio dilatation in the direction of acceleration"** [12, V2, D47, p. 302]. Contrary to this claim and according to length contraction fully accepted by Einstein after 1907, the shape of bodies are contracted with reliance on their velocity. But in physics there is no other possibility than for the acceleration to come from the one velocity to the other one.

In a 1911 paper [12, V3, D23] Einstein came to his physically utmost idiocy (here Feynman's simile is used), which he called "a consequence of fundamental significance" of his theory that "in the gravity field the frequency of light is everywhere the same but just the clock by which we measure the frequency runs slower"!

According to this Einstein's revelation of *"fundamental significance"* all our experimental observation of **various ratios of refractive indices** of the light (ratio of change of light frequencies) at the transition between different densities of the translucent substances or fluids **should always equal to one** and *"just the clock by which we measure the frequency runs slower"*.

Simultaneous validity of two opposite claims in STR represents even the very first postulate of STR about always constant velocity of light speed and primary claim of relativity that all velocity are relative.

Although allegedly biggest discovery of STR is that all velocities are relative and no absolute velocity exists, the STR is based on the absolute, fixed and never changing constant velocity of light. **So if the speed of light is the absolute velocity, then we can measure all other velocity as absolute taking the speed of light as base and the primary STR claim about relativity of all velocities is nonsense**.

On top of it in the second of two postulates of Einstein's special theory of relativity he simultaneously declares that all velocities of bodies to velocity of light are always the velocity of light. **So, if the speed of light is chosen as the basic comparative velocity then no other velocity than the velocity of light can be measured**.

On this schizoid base is constructed the special theory of relativity, allegedly the biggest achievement of the human spirit in the history of mankind.

Although the principle of non-impugn-able precise equivalence between the gravitational and inertial mass is according to all today's mainstream physics the fundamental base of Einstein's General theory of relativity Einstein renounced his equivalence principle. This can be best seen in Abraham paper of 1915 'Recent Theories of Gravitation', *"Einstein started from the equivalence hypothesis and derived the equations of motion of material points in static gravitational fields. Shortly thereafter, however, he convinced himself that in order to avoid contradictions with the law of conservation of momentum the equivalence hypothesis does not form a firm foundation for the theory of the gravitational field"*[65].

Einstein's own words in the paper quoted in this Abraham's paper are, *"I hesitate to take this step because by doing so I am leaving the territory of the unconditional equivalence principle. It seems that later can be maintained for infinitely small fields only"*.

In the 1913 paper [12, V4, D13, p.153] 'Outline of the generalized theory' we can read - *"I have shown in previous papers that the* *equivalence hypotheses leads to the consequence that **in a static gravitational field the velocity of light c depends on the gravitational potential. This led me to the view that the special theory of relativity provides only an approximation of reality**; it should apply only in the limit case where differences in the gravitational potential in the space-time region under consideration are not too great"*.

As the gravitational potential is changing in space from a star to a star, from a galaxy to a galaxy so, according to the 1913 paper, the velocity of light in a vacuum is no longer constant and is changing (standard supposition of physicist before 1905). **This means, in fact, an abolition of the first principle of STR which is based on the firm proclamation that the velocity of light is the ultimate and never changing constant velocity in the vacuum of void space, that no carrying substance of the light propagation exists and that nothing can influence the ultimate velocity of light in a vacuum.**

So the allegedly already precise STR laws from 1905 (contrary to Newton's and Maxwell's laws which are then allegedly just an approximation of STR laws) become in 1913 just approximations again and need further correction. This correction had been made yet in Einstein's 1912 papers [12, V4, D3, D4] where equation for STR momentum is corrected by multiplying it with difference of light velocity c and for a correct energy equation is stated equation $E = mc$ with alleged correct classical approximation $E = mc + mv^2 / 2c$.

The consequence of these Einstein claims that in the uniform gravitational field of the universe velocity of light is changing (as well as wave length in relativistic gravitation red-shift), is that Hubble's explanation of the law of red-shifts (as light loses energy in proportion to the distance it travels through space) is correct and the Big-Bang theory must be discarded.

Einstein's standard controversial procedure was also the case of so called central paper of GTR 'The Field equations of gravitation' from

1915 (4 pages) [12, V6, D25] **which was just another in a row of 15 published papers** by Einstein from 1911 till 1915 on this topic.

Yet Einstein's 14 page paper in 1912 ' On the theory of the static gravitational field' [12, V4, D4] ,which **contains no single reference, led to full controversy with Abraham**, who published four papers before Einstein in 1912 on gravitational fields in Physikalishe Zeitschrift Vol 13,1912.

Abraham accused Einstein of stealing the relation for the energy of the material point in the gravitational field from him as well as relations for the energy density and the stresses in a gravitational field, which Abraham had already given in his paper. In response to Abraham, Einstein in his paper [12, V4, D8] admitted his fault in the case of the energy of material point – *"I had, unfortunately, overlooked this"* and in the case of energy density and stress he disagreed with the strange argument - *"The way in which c enters in the two theories is different"*.

So when Einstein replicated Abraham's equations and, if Einstein opinion about role of light velocity c in these equations is different from Abrahams, then for Einstein it means that these equations are his and Abraham's authorship is redundant to quote.

In a similar uppish and arrogant manner Einstein in his paper [12, V4, D6, Response to a comment by J. Stark, 1912] is dealing with the requirement of Stark (Nobel Prize in 1919) to recognize his priority in the case of the law of photo-chemical equivalence – *"J. Stark has written a comment on a recently published paper of mine for the purpose of defending his intellectual property. I will not go into the question of priority that he has raised, because this would hardly interest anyone, all the more so because the law of photochemical equivalence is a self-evident consequence of the quantum hypothesis"*.

Nordström (in 1912) published a paper in which the gravitational field equation R= $\varkappa$ T was first derived, where R is fully contracted Riemann-Christoffel tensor, T the trace of the stress-energy tensor and $\varkappa$ = const. in the case of an unstressed, static matter distribution.

From a sum of the 15 papers published by Einstein from 1911 till 1915, the weighty one, Einstein's paper form 1913 'Outline of the GTR and of the theory of gravity' [12, V3, D13], **was filled with errors and its conclusions were incorrect**. Einstein's 1913 paper 'Outline...' as usual was very strongly criticized and labeled, mainly by Mie, **as incompetent**.

Einstein in his 1914 paper [12, V4, D25] 'On foundation of GTR and the theory of Gravitation' on 4 pages in 11 points tries to deal with Mie's objection and in the end admits "*I do not agree with the conclusion of this critique and I cannot escape the impression that Mr. Mie has not correctly understood my theoretical intentions. I believe that the incompleteness of my previous presentation of the main ideas of the theory is to be blamed for this misunderstanding*".

But afterwards Einstein corrected his 1913 paper in his 1914 paper 'Comments on Outline of the GTR and of the theory of gravity' [12, V4, D26] in which Einstein admits basic mistakes in his physical speculation and changed the former paper in 5 areas.

From 1915 till 1917, **at least 7 Einstein papers on GTR were published** with modifications and various explanations.

The so called Central paper of GTR from 1915 [12, V6, D25], was changed by Einstein in his 1916 [12, V6, D41] paper and in his 1917 paper [12, V6, D43] in which a cosmological constant was added to his field equations. But in 1927, Einstein abolished the cosmological constant, later calling it the 'biggest blunder' of his life (detailed below).

Yet in 1919, in the beginning of his paper [12, V7, D17] 'Do Gravitational Fields Play an Essential Role in the Structure of the Elementary Particles of Matter?', Einstein eliminated his cosmological constant as not necessary, condemned field equation discovered in his so-called Central paper of GTR from 1915 with the words "*Not only the problem of matter, but the cosmological problem as well, leads to doubt as to equation* $G_{\mu\nu} = -\kappa T_{\mu\nu}$" and he set forth another changed field equation. The result of his paper (contrary to more paper published at that time mainly by G. Mie which proved that matter is fully electric) is that matter (elementary particles) is constituted from three-quarters of electromagnetic field and one-quarter of the gravitational field – "*This equation signifies that of the energy constituting matter three-quarters is to be ascribed to the electromagnetic field, and one-quarter to the gravitational field*".

In a 1923 paper 'On the GTR' [12, V13, D417] Einstein proposed a another different equation of gravity fields than that from his so called central paper from 1915 and Einstein already regularly accepts ether filling the space- "***I assume a continuum in which a physically meaningful mass –invariant does exist…***" .

In the equations in Einstein's paper from 1917, **H. Thirring found mistakes in two relations** and, in a letter from December 1917, asked Einstein for explanation. "*How can these differences in the Coriolis force and centrifugal force be reconciled? The energy balance thus is not correct. There is nobody in Vienna who would be able to solve these paradoxes*" [12, V8, D401].

Einstein published his Field equation in a 1915 paper and said prior to its presentation that he, "completely succeeded in convincing Hilbert and Klein". Factually Einstein estreated the majority of his general relativity work from these two men as well as from Abraham and Nordström in earlier periods.

On November 16, 1915, Hilbert delivered a talk in Gotttingen presenting his new axiomatic derivation of the 'basic equation of physics'. Hilbert presented in this talk, given five days prior to Einstein, the correct, generally-covariant equation of gravitation that lies at the heart of the general theory of relativity. Einstein presented his paper on November 25, 1915 in Berlin.

Hilbert in his letter of 13 (or 14) Nov 1915 [12, V8, D140] to Einstein gave a brief sketch of his theory and claimed to have achieved a unification of gravitation and electromagnetism. He invited Einstein to his forthcoming presentation and discussion of his theory on 16 Nov 1915.

Einstein in a letter of 15 Nov [12, V8, D144] **excused himself for stomach pain and asked Hilbert for a copy of his talk**. *"I must refrain from traveling to Göttingen for the moment and rather must wait patiently until I can study your system from the printed article; for I am tired out and plagued with stomach pains besides. If possible, please send me a correcting exemplar of your study to mitigate my impatience"*.

Hilbert received a letter from Einstein dated on 18 Nov thanking him for sending a manuscript of the paper. This paper Hilbert delivered as the proofs version, dated on the 20 November to the Royal Society in Gottingen. So, in fact, Hilbert had sent a copy of his work at least a week in advance to Einstein before he delivered his lecture, but Einstein did not send Hilbert an advance copy of his own work [36].

Naturally this action by Einstein violated their friendly relationship, which is evident from the rich correspondence between Hilbert and Einstein before 25th November. That **Einstein was fully aware of his deceptive unfair action** is obvious from his letter to Hilbert from 20th December 1915 [12, V8, D167] in which Einstein asked Hilbert, to provide him pleasure and consider him a genius again - *"There has been a certain ill-feeling between us, the cause of which I do not want to analyze. I think of you again with unmixed geniality and **ask you to** *try to do the same with me. Objectively it is a shame **when two real fellows who have extricated themselves somewhat from this shabby world** do not afford each other mutual pleasure"*.

It is evident that, in this letter to Hilbert from 20 December 2015, Einstein acknowledges that Field equations are at least the work of both 'fellows' if they were not just the work of Hilbert alone.

Conclusions of Jagdish Mehra book "Einstein, Hilbert, and The Theory of Gravitation. Historical Origins of General Relativity Theory" [66] are that Hilbert was orimary creator of Einstein's Field equation from the 25th of November 1915.

Einstein several years before his presentation in 25. November 1915 put forth equations called "Entwurf equations". It is relevant to recognize [65, Michael Janssen] that mathematician M. Besso who was coworker on Einstein's Entwurf equations showed and taught Einstein physical reality that his approach based on Machian idea of gravitational potentials as an outcome of a rotation of the fixed stars is physically wrong due to revealing of Coriolis forces at decreasing diameter of rotation of a hollow or due to increasing inertial force with increasing acceleration in the void space.

When Einstein was acquainted with the results of Hilbert's work at the end of October 1915 he demised his approach in his Entwurf equations. His new field equations in paper of 25. November 1915 resp. in 1916 content the same number of gravitational and electromagnetic potentials as Hilbert's are derived fully from Hilbert's principles based on the variational principle of Hamiltonian of the word function of the ether [65].

These Einstein's new published equations were this time surprisingly generally-covariant field equations, but this general covariance was in full contradiction to his vigorous argumentation in previous years according to which generally-covariant field equations are physically unacceptable. Einstein offered no explanation, what was wrong with this argumentation that he had published four times before.

In Hilbert's March 1916 published paper of his presentation from 20 November 1915 Hilbert added *"It seems to me that the differential equations of gravitation so realized by me are in agreement with the beautiful theory of general relativity proposed by Einstein in his later [25 November 1915] memoir. But his (Einstein's) H functions are in no sense general invariants, nor do they contain the electric potentials."*

After this objection Einstein in his 1916 [12, V6, D41] paper corrected his equations from today so-called central paper of GTR from 1915, 'The Field equations of gravitation'.

Einstein's standard controversial procedure was also the case of Einstein's so called correct evaluation of the perihelion of Mercury - 'Explanation of the Perihelion Motion of Mercury from General Relativity Theory'- presented by Einstein at the Prussian Academy of Sciences in Berlin on 18 November, 1915 and published a week later as proceedings [12, V6, D24]. In Einstein's explanation he made two approximations, which means modification of physical laws of the two powers in two steps in the aggregate.

On 22 December, 1915 [12, V8, D169] Schwarzschild wrote a critical letter to Einstein in which he criticized Einstein for mistakes in the successive approximation approach and **he required correction of his paper**. In his letter Schwarzschild penned *"In order to be able to verify your gravitational theory, I have brought myself nearer to your work on the perihelion of Mercury, and occupied myself with the problem solved with the First Approximation. **Thereby, I found myself in a state of great confusion. I found for the first approximation of the coefficient gμv other than your solution... A not too difficult calculation gave the following result**... "*.Schwarzschild letter ends

*"As you see, it means that the friendly war with me allows this stroll **in your fantasy land**".*

Since then Einstein never returned to the solution of the GR perihelion advance problem or to light bending in a gravitational field. This means that Einstein never solved allegedly 'his own' GTR Field equation as he recognized that he was an incompetent amateur in these topics.

Among the key issues of GTR, the problems of energy and angular momentum conservation, along with the properties of stress-energy momentum tensor, remain the 'hot' (better say, controversial) ones. Possibly, this is one of the reasons why there are numerous publications devoted to the GTR perihelion advance effect and light bending, which are considered controversial or arguable – *"**A fatal mistake in Einstein's final solution in the problem of planetary perihelion advance is the assumption that the impact of the GR term on the roots x1 and x2 is negligible**"* [37].

Perhaps the best comprehensive analyses of Mercury's perihelion and bending light around Sun is the work of Anatoli Andrei Vankov [38]. Conclusions of this work are:

"An introduction of GR relativistic effects inevitably leads to complications of the problem formulation for, at least, two reasons: a) such quantities as potential and kinetic energies are out of the GR arsenal; b) the exact GR N-body solution, which would have the Newtonian limit, does not exist.

The GR effect of the Mercury's perihelion advance is claimed to be successfully confirmed. In the present work, the results of our analysis of the problem show that the claimed confirmation is not true.

The GR-term, which is thought to be the cause of both the perihelion advance effect and the light bending effect, has no physical sense, as shown by Fock and in our work. Fock's work also shows that the equation of light propagation in a vicinity of massive object must have the linear (potential) term instead of the GR-term.

Controversies in the GR theory arise in connection with gravitational properties of both the particle and the photon. *The GR prediction of the particle deceleration in a radial free fall makes no physical sense and most likely is wrong. The prediction of the bending of light is not valid for the same reason as in the perihelion advance case, while the predicted red-shift is inconsistent with the GR framework. These and other arguments raise the question about the sufficiency and completeness of the GR physical foundations. Overall, we conclude that the claimed confirmation of the GR prediction of the relativistic perihelion advance is neither theoretically nor empirically substantiated."*

Another situation in the row of simultaneous validity of two exclusive opposite claims in STR and GTR is the situation with so called cosmological constant.

Einstein's field equations of 'genius' from so called central paper of GTR from 1915 [12, V6, D25] (4 pages) guaranteed as correct and glorified to heaven by academics and mass media to the present days as 'the greatest achievement in the history of mankind', **resulted in a rapid gravitationally collapsing universe. So, soon after Einstein concluded that his Field equations were not utilizable to the Universe. In order to save his 1915 field equations from inevitable rapid collapse**, Einstein on 11 pages in 1917 paper [12, V6, D43] substantially changed the equation form 1915 by **including another term in the equations**, the **so called cosmological constant**, thereby keeping the universe steady.

61

So the claim of special relativity from 1905 and general relativity from 1915, that no ether exists and space is totally void, is in 1917 given the opposite claim of its creator in a cosmological constant **that space energy density of the vacuum--so ether in space exists**.

This expression, 'energy density of the vacuum', is clearly logical nonsense which brings the standard 'non ingenious' physicist to a schizoid state. Either there is a vacuum or there is energy, so mass as it was 'revealed' by Einstein in 1923 is a *"physically meaningful mass invariant"*.

The same expression 'energy density of the vacuum' is also used by Schrödinger in quantum mechanics and Heisenberg's quantum mechanics coined the words "vacuum fluctuation" or "Zero-Point fluctuation" to take care of that problem. Expression fluctuation of zero-point energy is similarly clear logical nonsense. Either there is a vacuum, so zero energy or there is a non-zero point of energy.

As energy is equivalent to mass this means that all space is filled with matter.

Einstein's cosmological constant (the energy density of vacuum, so substance, matter or ether) from 1917 had properties of inactive substance that cause a slowing down in the otherwise rapid gravitational collapse of the Universe resulting from Einstein's 1915 Field equations.

This claim and of ether's 'new rediscovery' by the genius was, of cause, not glorified by the ideological power structures. But, as from 1917, it becomes more and more clear that Andormeda nebula is independent galaxy, the rotation of which is not governed by Newton's law. Einstein was abandoning the cosmological constant in his paper [12, V7, D17], [39] from 1919, calling it *"gravely detrimental"* and eliminated his cosmological constant *"as not necessary"*.

A. Friedmann noted in 1922, mistakes in Einstein's solution of Field equations with the included cosmological constant in 1917 and concluded that the field equations from 1915 are sufficient for many types of Universes and that introducing the cosmological constant was an error. Einstein very quickly admitted Friedmann's claim and published his remarks announcing that *"the cosmological term is now not necessary"* [12, V13, D340].

Afterwards, when an expanding instead of a static universe was proposed by priest Lemaitre in 1927, Einstein renounced the cosmological term in 1932, in a joint paper with de Sitter [40].

Later Einstein continuously regretted the introduction of the cosmological constant, calling it the "biggest blunder" of his life [41, George Gamow].

Thus the Cosmological constant was repudiated by Einstein, academics and mainstream physics till 1990, when the Hubble Space Telescope was launched. The Hubble Space Telescope unveiled new regions of the Universe to us and academics again reintroduced Einstein's cosmological constant. This, despite Einstein definitely repealed it calling it the biggest blunder of his life and despite Freedman showing that it is redundant to receive many types of Universes from Field equations.

The cosmological constant, included again into Field equations after 1990, is allegedly the cause of the expanding universe. This constant has the function of fake up force that stems from so called dark energy and is allegedly counteracting gravity. The cosmological constant now has a bizarre physical interpretation conjured in the Universe as generally valid, everywhere constant, repulsive force that increases with increasing distances between two objects and depends only upon the mass of one of them!

But what would we not do for the salvation of the universe?

But, as is detailed below in chapter 4. the claim of contemporary mainstream physics about everywhere in space existing, repulsive force so everywhere existing dark energy also evidences that space in the Universe is not empty, but is filed with dark energy; therefore with matter and therefore with ether. This is, however, the evidence that STR and GTR, based on the nonexistence of ether, are false and must be rejected.

The regular controversial procedure, when Einstein declared both of these mutually opposite excluding claims, was **also the case in question of the existence of the ether**.

Einstein discarded the existence of the ether in 1905 and in his 1914 paper "On the principle of relativity" [12, V6, D1] he still indicates abandoning an ether as the main result of relativity - *"Of the major results of the theory of relativity two should be mentioned here. First: the hypothesis of the existence of a space-filling medium for light propagation, the so-called light-ether, has to be abandoned. Light appears according to this theory no longer as a state of motion of an unknown carrier, but rather as a physical structure with a physical existence of its own"*.

But yet in 1916 paper he regretted that he rejected it and speaks about the introduction of a medium filing the space and assumes that electromagnetic fields are ether states-- *"metric facts can no longer be separated from true physical facts; the concepts of space and ether merge together. It would have been more correct if I had limited myself, in my earlier publications, to emphasizing only the non-existence of an ether velocity, instead of arguing the total non-existence of the ether"* [42].

This new position and acceptance of ether was, as usual, not Einstein's idea and, as usual, he did not mention his source of this idea. Einstein's new concept of the ether was born out of objection on rejecting the ether in works of H.A. Lorentz, Ph. Lenard, H. Weyl, G. Mie, D. Hilbert in 1916 and 1917 and their personal letters to him [42].

These physicists (mainly G. Mie) in their works in this period were deducing the creation of particles and field equations (gravity) form homogeneous electromagnetic field, including the electric charge density, the convection current and the magnetic field strength in interstellar space without the necessity of existence of other forces as the gravitational forces or others. As a reaction to these works Einstein wrote in 1919 paper that matter (elementary particles) also according to his General theory of relativity is constituted from electromagnetic but also from the gravitational field (as was detailed above).

Einstein in his works from 1905 till 1907 discarded the ether from physics, but his more than 5 papers from 1920 to 1934 covered the matter of ether as an unexceptionable physical reality. In his published lecture at a conference held in Leiden in 1920 [43], in his another paper published in his 1920 [12, V7, D31], and especially in his work 'On the ether' from 1924 [12, V14, D332] **Einstein declared the opposite claim that, without the ether, it is not possible to explain the physical world around us**.

Einstein concluded in his lecture in 1920 [43] –"*Thus we may also say, I think, that the ether of the general theory of relativity is the outcome of the Lorentzian ether, through relativisation. According to our present concepts the elementary particles of matter are also, in their essence, nothing other than condensations of the electromagnetic field. According to the general theory of relativity, space without ether is unthinkable. ...ether has to serve as a medium for the effects of inertia. Recapitulating, we may say that, according to the general theory of relativity, space is endowed with physical qualities; in this sense, therefore, there exists an ether.*"

Einstein, in a 36 page 1920 paper, '*Fundamental ideas and methods of the Theory of relativity, presented in their development*' [12, V7,

D31], **which do not include any single reference**, declared his error in rejecting ether in 1905:

"My opinion in 1905 was that one should no longer talk about the ether in physics. But this judgment was too radical. Again, 'empty' space seems to be endowed with physical properties, that is, not physically empty as it appeared in the special theory of relativity.

Rather it is still permissible to assume a space-filling medium whose states may be imagined as electromagnetic fields. Therefore, one can say the ether has been resurrected in the theory of general relativity. Space and ether flows into each other…The theory of space (geometry) and time no longer represent intrinsic physics propounded independently of mechanics and gravitation".

This meant the refutation of STR by Einstein himself.

In the conclusions of Einstein's 1920 papers [12, V7, D31] Einstein rediscovered the ether. He apparently newly 'discovered' the ether as the direct consequence of 'his' GTR field equations, although full assurance of the ether by physicists and hundreds of papers in topics of ether from Aristotle until 1905 existed.

In the 1924 paper (when continuous Millers experiments kept confirming ether) Einstein became more of an enthusiastic advocate of the existence of the ether than supporters of the ether before the year 1905. In his 1924 paper Einstein 'discovered':

– *"The mechanical ether, designated by Newton as 'absolute space', must therefore be considered by us as a physical reality".*

- *"In Newton's theory of motion, space has physical reality- in contrast to the case of geometry and kinematics".*

- *"GTR adds characteristics to the ether that are variable from point to point and determine the metric and the dynamic behavior of material points"*

-*"The ether of the GTR, consequently, differs from the one of* **classical mechanics, i.e. the special theory of relativity**, *in that it is not 'absolute'; its local variable properties are rather determined by the ponderable matter".*

- " **we are not going to be able to dispense with the ether in theoretical physics , that is, with continuum furnished with physical properties;** *because GTR excludes any unmediated action-at-a-distance. However, every theory of contact action presupposes continuous fields, hence also the existence of an ether".*

But after these declarations in his 1924 paper and according to his own affirmation in 1925, his STR and GTR become invalid. "***My opinion about Miller's experiments is the following. Should the positive result (ether) be confirmed, then the special theory of relativity and with it the general theory of relativity, in its current form, would be invalid***" [25].

Till 1905 many physicists worked on several central topics of the physics of that time as the energy contained in the mass, the energy of the light quanta, the photoelectric effect, the contraction of bodies moving in the ether. They carefully formulated feasible physical rules of observed phenomena rather as the proportionality, then the physical law before these rules were experimentally and contextually confirmed.

In Whittaker's book [46] an important chapter entitled "*The relativity theory of Poincaré and Lorentz*", these two authors are presented as the incontestable creators of relativity. Einstein's own contributions are restricted to the discovery of the formulas of aberration and the Doppler effect.

Poincaré from 1899 till 1904 (the lectures at Sorbonne, proceeding from conferences in Paris, Sant Luis) declares that absolute motion cannot be find on dynamical, optical or electromagnetic bases. **The general form of this principle Poincaré called "*the principle of*** *relativity"* **on 24 September 1904 at the Congress in Saint Luis. He declared that** *"in accordance with the principle of relativity the physical law has to be the same as for observer in rest as well as for observer in uniform translation motion. From this principle a new sort of mechanics must arise in which no one velocity can exceed the velocity of light"*.

E. Giannetto in his 1998 comprehensive work [51] 'The rise of Special Relativity: Henri Poincaré's works before Einstein' concluded that there is no doubt **Poincaré must be considered the actual creator of special relativity."**

All these basic notions of Poincaré were stolen by Einstein into STR without proper references.

Lorentz and Larmor were working on this new sort of mechanics from 1900 till 1904. They developed the basic equations retreated by STR and Lorentz also introduced a notion of local time.

Lorentz in 1909 perseveringly explained his transformation by the existence of the ether and his opinion about Einstein's and Minkowski's relativity was that one can come to the same results if one deny the existence of the ether and of true time and space, and to see all reference systems as equally valid.

In a paper at the French academy of science published on 5. June 1905, Poincaré crowned the logic of Lorentz about covariant transformation of Maxwell equations to Lorentz transformation and received the equation for the transformation of density of electric charge end current. Einstein on 30 June 1905 delivered his paper for publication 'On the electrodynamics of moving bodies' [12, V2, D23]. Einstein's 1905 paper, a so called central work of special relativity was his first paper on this topic. **The paper contained not a single** reference although, as was shown, at least 10 papers with similar content were known at that time.

In Einstein's paper we can trace all notion and equations to those of Lorentz, Poincaré and others. In this paper Einstein speaks about this notion and equation as *his* discoveries. For example, the following sentence from Einstein's 1905 paper is punctually a declaration of Poincaré from 1904, as mentioned above:

"We will raise this conjecture, the purport of which will hereafter be called the 'Principle of Relativity', to the status of a postulate, and also introduce another postulate, which is only apparently irreconcilable with the former, namely, that light is always propagated in empty space with a definite velocity c which is independent of the state of motion of the emitting body".

H. Minkowski in his paper Relativity principle, published in 1909, declared that Einstein was by no means the principal expositor of the principle of relativity and penned about the Lorentz paper from 1895 mentioned above *" H. A. Lorentz has found out the 'Relativity theorem' and has created the Relativity postulate. "*

More proposition of proportionality $E \approx mc^2$ was propounded till 1905 e.g. Nikolay Umov - en.wikipedia.org/wiki/ Nikolay_Umov *"He was the first scientist to indicate an interrelation between mass and energy proposing the formula $E = kmc^2$ as early as in 1873".*

In 1675 Newton had already argued that light is composed of particles and posits the existence of the ether to transmit forces between the particles.

The word 'quantum' was in general well known before 1900. Riemann in 1854 called Quanta the discrete portions, little hills on a surface of flat continuous manifold and word quanta was used for energy of heat by Helmholtz and Mayer after 1870. Boltzmann had used the notion of energy quantization in his research publication as early as 1872, in which he divided the energy of a system into extremely small, discrete packages. Boltzmann in 1877 suggested that the energy levels of a physical system could be discrete.

The concept of quantization of radiation was used in 1900 by Max Planck assuming that energy of radiation can only be absorbed or released in tiny, differential, discrete packets he called "*bundles*" or "*energy elements*". Hertz, Thompson, Stark and Planck came to believe that light is formed in quanta and Planck expressed the relation for energy of quanta as $E \approx h\nu$.

The photoelectric effect was fully experimentally and theoretically explored by H. Hertz in 1887, J.J. Thompson in 1898 and, above all, by Lenard in a 1902 paper [59] after ten years experimental and theoretical study. Lenard proved that in photoelectric effect the maximum kinetic energy of released electrons is determined by the frequency of the light so that $E_{max} = m\nu^2 / 2 \approx k\nu$.

Around 1890 Lorentz and Fitzgerald propounded a contraction of dimensions of solid bodies $l = l_0 \sqrt{1 - \nu^2 / c^2}$ as a consequence of the different pressure of the ether at different speeds of bodies within the ether in proportionality to change of energy $\Delta E \approx \nu^2 / c^2$.

The gravitational weakening of light from high-gravity stars was predicted by Michell in 1783 and Laplace in 1796. The effect of gravity on light was then explored by Soldner (1801), who calculated the amount of deflection of a light ray by the sun. All of this early work assumed that light could slow down and fall.

Einstein acted fully in accordance with the dictum in which he proclaimed that 'genius is in knowing how to hide your sources'. He simply usurped the results of these physicists and did not cite authors of these ideas and their results in his works. In every other field of intellectual property, except science, he would have been sued for this procedure in the courts.

Proportionality as the first estimation of results of these physicists, which required further deep experimental study Einstein simply wrongly replaced by equality and discovered 'physical law'. But worst for physics is that his unproven declaration of physical law (he called his 'discovery' as heuristic) must for him had have the linear form needed for his linear contemplation between the difference in run off distance and change in velocity at Galileo's transformation in his STR structure $x' = x - vt$.

Einstein's first postulate is that "*the laws of physics undergo no change and are the same in frames with uniform translation motions*". But there is no physical law in physics concerning of uniform translation motion except Galilei's and Newton's constatation (called Newton's first law) that all bodies remain in rest or uniform translation motion if no forces act on them. This constatation means that, from the point of view of physics, nothing is changed when bodies remain in the same physically stationary states that could be survived by Kinematics and no physical law is related to these states.

Einstein in his STR could be entitled to a produced transformation relation (subsisting in different measurement results of lengths, time and velocities) between frames at different velocities with account to constant velocity of light, so transformation based on his two postulates of STR. But physics (kinematics) based on these two postulates are concerns of no physical laws and so exclude a priory to derive any physical law in STR or correct any existing physical law from it. But STR commanded that all physical law must obey these transformations that flow from these two basic postulates and bring a number of new physical laws into so called relativistic mechanics.

In this way, the physical laws of classical mechanics (dynamics) were converted by Einstein into Einstein's law of kinematics.

Physical laws (dynamics) are derived from physical processes when the rest states or uniform motion of bodies is changed. So the physical laws concern those situations when Einstein's inertial frames are accelerated in non-linear ratio of space and time to different velocities so they do not concern STR. This is, as was mentioned above, what Einstein finally fully understood just after 1920: "*If we had based our considerations on the Galilei transformation we should not have obtained a contraction of the rod as a consequence of its motion*". This, his own declaration, represents a rejection of special relativistic mechanics.

Einstein, in a 1907 paper, proclaimed that Newton's law of motion must be the same in uniform translational motion and this proclamation he called the principle of relativity. But Newton's law $x = x_0 + v\Delta t + a\Delta t^2 / 2$ includes acceleration (which produces kinetic energy $mv^2 / 2$) and so evidently cannot be linearly transformed by Galileo's transformation because different reference frames are shifted by the different quadratically growing adding constant of kinetic energy to different frames. This means that different inertial frames are not equivalent as is claimed in his first postulate of relativity. So Einstein simply proclaimed the fraud that classic kinetic energy is incorrect because it is (naturally) not conserved (linearly) at linearly increasing velocity at Galileo's transformation. He faked his "correct" new kinetic energy E=mvc=pc and classical kinetic energy was degraded as the teleological fraudulent approximation of Einstein's new faked kinetic energy.

So he based his physics in STR and QM on his new faked energy system $E = mc^2 = h\nu = h\nu / c \cdot c = h / \lambda \cdot c = pc = mvc$ without any physical proof, reasoning or connection to the previous development of physics. All terms in these equations consequently represented basic physical law in the 20th century, but Einstein did not provide any explanation or definition of what these terms meant. But these equations, 'Einstein's great discovery', are in contradiction with all experimental physical laws of all other parts of physics except STR.

In all parts of physics, energy equals to a quadratic relation of these parameters. Energy equals to quadratic relation with frequency $E \approx \omega^2$ in classical mechanics, quantum mechanics and electrodynamics. Energy of source in electrostatic and gravistatics equals $E \approx Q^2 \approx m^2$ and energy in particle physics $E \approx p^2 \approx m^2 v^2$ what for Einstein is $E^2 \approx \omega^2 \approx p^2 \approx m^2$. In quantum physics and quantum mechanics energy is proportional to $E \approx 1/\lambda^2$ while for Einstein it is $E \approx 1/\lambda$. Energy to velocity, time and distance is in quadratic relation in classical mechanics $E \approx v^2 \approx a^2 t^2 \approx s^2$ what for Einstein in STR is $E \approx v \approx at \approx s$.

All quadratic relations above express the fact that Dynamics, contrary to Kinematics, is nonlinear. That means that for changing of a physical state, the difference in the density of the matter so gradient of force fields (described by nonlinear ratios of space and time so change in velocities so by accelerations so forces) must exist.

Although one of the main results of Einstein's Special theory of relativity (interim 1905 - 1910) is the equation for energy $E = mc^2$ with alleged classical approximation $E = mc^2 + mv^2 / 2c^2$, celebrated around the world for next hundred years up to now as allegedly the greatest discovery of mankind, Einstein already in 1912 in his paper [12, V4, D3] brings another 'great mathematical discovery' of this physical law stating that correct form of these equations for energy are $E = mc$ with alleged classical approximation $E = mc + mv^2 / 2c$. This time these equations do not even have the dimension of energy. Later in the

General theory of relativity (e.g. in Einstein's energy-momentum tensor) this equation has the form $E = m^2c^2$ or $E = m^2v^2$ or $E = m^2vc$ but not the linear form $E = mc^2$.

Definitely relation between energy and mass with reliance on the velocity or with reliance on accumulated forces in matter or with reliance on strength of force fields around matter cannot be linear and Einstein's fake relation $E = mc^2$ is a blunder.

Planck in a 1908 paper stated that this Einstein result *"is permissible only in the first approximation"* [26].

Einstein himself later as well understood that this equation is only an approximation. In his paper " $E = mc^2$: The most urgent Problem of our Time" [Sci. Illustr., I, 1946] he announced *"It is customary to express the equivalence of mass and energy (**though somewhat inexactly**) by the formula E=mc²"*.

The result of everything said above is that Field equations can hardly be considered as Einstein's, because he was the last who understood these equations, was not able work with them or expertly solve them.

Hilbert in 1915 quickly understood Einstein's capabilities at this point. In his lectures he stated, *"Every boy in the streets of Gottingen knows more about 4-dimensional geometry than Einstein"* or, *"Do you know why Einstein said the most original and profound things about space and time in our generation? Because he learned nothing at all about the philosophy and mathematics of time and space"* [66].

To this Einstein remarked, *"The people in Gottingen sometimes strike me, not as if they want to help one formulate something clearly, but as if they only want to show us physicists how much brighter they are than we"*.

As was shown above, Einstein was a fully incompetent amateur among professional mathematicians (Hilbert, Schwarzschild, Friedmann, Levi-Civita, Grossmann, Besso) at this level of mathematics.

Einstein was fully aware of this fact, he accepted their critique and did not oppose their refutation of the way he worked with Field equations or refutation of his Field equation solutions. **Einstein never performed an accurate solution of his alleged GTR equations for the only two cases for which he tried to solve them. The perihelion of Mercury and bending of light around the Sun are falsely presented to publicity as his great success** and decisive results allegedly confirming the validity of GTR.

As is shown above, Einstein was an incompetent amateur also in physics for all the then most outstanding professional physicists (Lenard, Mie, Abraham, Boltzmann, Laue, Stark) possibly with the exception of Planck and Lorentz in some cases. Einstein tried to dispute, oppose or object to these physicists, but at the end he mostly accepted their opinions.

We must ask why weren't dozens of Einstein's papers peer-reviewed before they were published in the professional journals by these professionals, who then were forced to publish works refuting these works of Einstein or asking their rejection in other ways?

If Einstein's papers produced behind the desk of patent clerk were peer-reviewed by professionals the status of one hundred years lasting relativity and quantum mechanics controversies could be avoid. Tens of needless Einstein correction papers as well as disapproving papers of professionals supplying peer-reviewing practice showing mistakes in Einstein's papers could be avoid.

But the breakage of physics into many separate parts using non-physical mysteries of relativity, quantum mechanics and Standard Model was welcomed be ideological power structures.

Einstein had insufficient education (four year of Polytechnic school) and insufficient experimental experience (at Polytechnic school he failed in a two year physics course Physical experiments for beginners) for the work as a theoretical physicist that he was trying to achieve in his life.

This book shows that Einstein did not really understand basic physical scientific principles and that he was writing about a topic that he did not fully understand.

Across all Einstein's paper the basic physical law and declaration of Galilei, Newton, Maxwell and other are never cited and are introduced countless time differently, deceitfully or deformed.

As an example we can bring out the Einstein's statement which is evidently fraudulent. Newton's quoted factual more opinions in which Newton resolutely rejected the action at the distance without ether medium *"an etherial substance, capable of contraction and dilatation, strongly elastic"* are offered in chapter 1. and 2. of this book.

These facts Einstein falsified in paper [43] *"The success of the Faraday-Maxwell interpretation of electromagnetic action at a distance resulted in physicists becoming convinced that there are no such things as **instantaneous actions at a distance (not involving an intermediary medium) of the type of Newton's law of gravitation**"* or in paper [12, V14, D332]: *"**It was Newton's theory of gravitation that first assigned a cause for gravity by interpreting it as action at a distance**. And yet this theory evoked a lively sense of discomfort among Newton's contemporaries, because it seemed to be in conflict with the principle springing from the rest of experience, that there can be reciprocal -*

action only through contact, and not through immediate action at a distance."

Across all Einstein's paper after 1920 we can trace the statement that ether was envisaged before him as the static immovable medium and Einstein's GTR now bring his new concept of this medium as dynamic substance. All physicist before Einstein from the Newton times envisaged ether as dynamic, elastic, rotating and swirling medium.

How far goes Einstein's plagiarism (here Lenard similar is used) and obsession that he discovered all physical low in physics we can see in [12, V6, D30, §13]

*"A freely movable body not subjected to external forces moves, **according to the special theory of relativity**, in a straight line and uniformly."*

Punctually this statement is learned each student on the first lesson of physics as Galilei's and Newton's law (called as Newton first law) - 'All bodies remains in rest or uniform rectilinear motion if no forces act on bodies'.

This is precisely the same case as is mentioned above when Einstein rediscovered the ether and in conclusions of his 1920 paper he claims *"**the existence of the ether is the direct consequence of GTR field equations**"* although full assurance of the ether by physicists and hundreds of papers in topics of ether from Aristotle ages till 1905 existed.

Note: Definition of "to plagiarise"- Webster's Dictionary – "To steal or purloin and pass off as one's own (the ideas, words, artistic productions, etc. of one another); to use without due credit the ideas, expressions or productions of another".

In this case we can point at the comprehensive 408 page book by C. J. Bjerknes, '***Albert Einstein, the Incorrigible Plagiarist***', 2002 [54], and its main conclusions : "***When we actually examine the life of***

Albert Einstein, we find that his only 'brilliance' was in his ability to plagiarize and steal other people's ideas, passing them off as his own".

Einstein piled up one mathematical construction after another with the goal to obtain predetermined results. Einstein, as a third-class patent clerk (1902-1909) and after four years non high-school study on a teacher's diploma produced reckless theories in all areas that were the main areas of interest to physics. He produced all these reckless theories during several years when other world top-rank physicists (Hertz, Thompson, Planck, Lorentz, Compton, Boltzmann, Lenard, Wien) had already been working experimentally and theoretically over several decades in only one of these areas.

Whenever an interesting theoretical paper or experimental result was published, Einstein immediately published his own theory to this published new theory or experimental result and it was of only minor importance for him if his theories were physically correct or not. Many times, Einstein in his theories simply confirmed theories and experiments of these world top class physicists from his hazy (heuristic) mathematical point of view.

By this acting these theories and experiments became common theories of Einstein's and of these world top rank physicists. He degraded the physics of the 20th century in many theories, as when he simply changed the proportional relations of these world top rank physicists to positions of equality.

These specialists had their own theories to their experiments and no one longed for theories from Einstein as they simply created one problem more in refuting Einstein's theory, as was first the case with his dissertation work detailed above. This naturally caused conflict and a wave of opposition from these specialists. Einstein, according to the objections of specialists from each of these areas, subsequently substantially repaired his primordial theory in tens of following corrected published papers.

Einstein, from the beginning of his career till its end, nominated himself to the position in which his task was to fake up theories and the role of others was to repair and to correct his theories, as well as teach him how "his" theories could be right and carry out the required experiments confirmed, repaired or refuted always "his" theories.

It is feasible to mention the opinion of the outstanding physicist, Max Born (Nobel prize award in 1954): *"Basic Einstein's work contains not a single reference to previous literature. It gives you the impression of quite a new venture. But that is, of course, as I have tried to explain, not true."*

Einstein's theories are proclaimed as his theories when he simply replaced the cautious claims of these specialists about proportionality by a sign for equality. Einstein's theories are also proclaimed as his, even when specialists in the essential parts substantially repaired his theories. Einstein's theoretical works are then characterized by publishing inceptive theories, followed by publishing many corrections, reparations and amendments to these inceptive theories. Most of Einstein's theories Einstein himself repaired, retracted or corrected according the experimental and theoretical work of other physicists.

This dialectical procedure, in which Einstein and contemporary relativity offer allegedly primary valid physical statements and, at the same time, also claim as valid the opposite claim to these primary statements is repeated many times in all special and general relativity. No one knows what relativity factually says and so these statements actually cannot be refuted or confirmed. The result is a conflict of frenzied circular patterns of thought and action, confusion and destruction of logical, philosophical thinking of man and the destruction of physics as a whole.

On such a schizoid basis is constructed the Theory of Relativity, allegedly the greatest achievement of the human spirit in all the history of mankind. This schizoid basis is, in relativity, incorporated into mathematical constructions of the highest level of mathematical difficulty of magic covariant and contravariant tensors which pretend as if they were the highly learned truth.

As was shown above, Einstein by his own declaration in 1920 or 1924 papers about the existence of the ether as an unexceptionable physical reality, in fact openly canceled his Special and General theories of relativity himself.

In spite of this fact, up to today, during the next 90 years to students and the general public it was claimed by ideological power structures, academics and main stream physics that STR proved the non-existence of ether and that GTR in a new understanding of space and time discovered the biggest achievement in the history of mankind.

Only the M-M experiment from 1887 is referred to in this claim, although over the next forty years continuing M-M experiments confirmed the ether and motion of the earth as determined by the ether-drift (detailed below).

In this claim just the Einstein's STR paper from 1905 is referred to, although Einstein himself, as is documented in this book, later canceled the basic principles of STR from 1905.

In this claim just the Einstein GTR paper from 1915 is referred to although Einstein (as is documented above in this book) soon concluded that his Field equations resulted in a rapidly gravitationally collapsing universe and was not utilizable to the Universe.

In this claim of discovery of a new understanding of space and time just the Einstein's GTR paper from 1915 is referred to, although Einstein fully rejected his space-time concept and fully returned to the ether concept after 1920.

4. Logical and Physical Distortions of Special and General relativity

Einstein's GTR and all theories of the gravity over past four hundred years are not about the curvature of space and time, but are about the shrinkage of the ether.

All theories of gravitation over the past four hundred years from W. Gilbert (from around 1600) through Descartes, Newton, Gauss, Riemann, Lorentz, Mie, Minkowski, Nordstrom, Abraham, Einstein to Hilbert (till around 1916) are based on two main basic physical principles.

Firstly, that flat continuous manifold of space is filled with matter (ether) with uniform density ρ described in these theories by world function marked as H or Φ modeled by Hamiltonian, Lagrangeian or field potentials.

Secondly, that shrinkage of this uniform space density of the continuous field of matter resulting in a change of the density ρ generates whole mass objects and/or mass particles of matter. These mass objects in return generate stress on uniform space distribution of matter (ether) in their surroundings and thus generate electromagnetic and/or gravitational potentials or, generally speaking, force fields.

In all these theories the real material homogenous continuous field of ether which fill in the whole Universe is a storehouse of mechanical momentum pressure, which reveals itself around great density areas (mass objects) as the consequence of impression of these areas on the homogenous field ether and execute impetus on bodies placed in them.

Gilbert (1600) declared that the gravity of the earth is of the magnetic nature which stems from electric and magnetic forces of earth elements.

Riemann (1861) came up from space filled with the ether whose compression and orientation resisting elements (Quanta), when compressed, produce scalar and vector potentials of electric, magnetic and gravitational forces. Riemann in 1854, *"space in itself is nothing more than a three-dimensional manifold devoid of all forms ; it acquires a definite form only through the advent of the material content filling it and determining its metric relations"* [5, p. 98]. The particles of matter which Riemann called Quanta represented according to him the discrete portions, little hills, on a surface of a flat continuous manifold of space filled with matter.

Lorentz and Mie (who called it the ether physics) at the beginning of the 20th century on the bases of their lifetime experimental and theoretical research of electrons took the particle as the states of ether created from ether's world function H. In their theories, they exploited action between particles of matter and ether which generates electrodynamic potentials, electromagnetic fields, currents and gravitational potential as a remnant of atoms shielded by the ether.

To this end, Lorentz proposed his transformations as a contraction of particles under the pressure of static ether relative to the particles which are a source of potentials of ether. If particles or bodies in their prime reference frame states obtain kinetic energy they contract according to the Lorentz transformation to the resulting energetic state. Emitting their energy into the ether they return to prime states. The covariance for Lorentz meant that different energetic states are not invariants and are linked with his Lorentz transformations.

Einstein changed this reality, introduced so-called inverse Lorentz transformation and announced that all reference frames regardless of their different velocities so regardless of their different energies are equivalents. He declared that the basic physical realities are invariants. Einstein's covariance means that all differences between different energetic states are just a consequence of different results in measuring of times, distances and velocities. All inertial and non-inertial frames are equivalent invariants and covariance means that there are no changes at all and all changes are absorbed in his math of kinematic.

The gravity, according experimental physicists Lorentz and Mie is a consequence of electrodynamics as e.g. in Lorentz's paper from 1900, 'Considerations on Gravitation', *"As to the electromagnetic disturbances in the aether which might possibly be the cause of gravitation, they must at all events be of such a nature, that they are capable of penetrating all ponderable bodies without appreciably diminishing in intensity. Now, electric vibrations of extremely small wave-length possess this property;* "

H. Minkowski (primarily mathematician) in his space-time fully acknowledged ether and exploited ratio of electrodynamic to electrostatic units and introduced rotation transformation for Maxwell magnetic.

Minkowski lecture given at the 80th Meeting of the Natural Scientists in Cologne on September 21st ,1908 - *"I want to make it quite clear what the value of c will be with which we will be finally dealing. c is the velocity of the propagation of light in empty space. To speak neither of space nor of emptiness, we can identify this magnitude with the ratio of the electromagnetic to the electrostatic unit of the quantity of electricity"* [67].

Or in 'The Relativity Principle', lecture given at the meeting of the Göttingen Mathematical Society on November 5th, 1907 - *"Now without recognizing any hypothesis about the connection between ' Ather ' and matter,..."* [67].

Abraham exploited Lorentz and Minkowski works and used four-dimensional tensor that should be considered in electrodynamics with ten components – the six components of electromagnetic pressure, and the three components of the energy current and the electromagnetic energy density.

Hilbert in his axiomatic approach (paper The Foundation of Physics from 20. November 1915) as the first axiom used Mie's world function H. He proposed a variational argument for modeling a 'world function' H, depending upon Hilbert's speculative physical fantasies of existence of 10 gravitational potentials, their first and second derivatives, the 4 electromagnetic potentials, and their first derivatives. His second axiom was that the world function H is an invariant with respect to arbitrary transformations of the 'world parameters' which he used instead of Einstein's space-time. However, as only from 10 gravitational potentials with their first and second derivatives can be deduced 4 electromagnetic potentials with their first derivatives and not vice versa Hilbert claimed that electrodynamic is an effect of gravitation, "*we can immediately assert the claim that in the sense explained above electrodynamic phenomena are effects of gravitation*". Hilbert introduced 10 gravitational potentials also by reason of providing mathematical tools to deal with the general covariance (relativity) principle.

The evaluation of Hilbert's fantasies about the existence of 10 gravitational potentials best reflects a critique of M. Abraham in his paper from 1915 'Recent Theories of Gravitation', "*the supposition, that there exist ten gravitational forces instead of one, is not supported by experience.*"

Einstein, in previous years until the presentation of his work 'The Field equations' in 25. November 1915, put forth equations labeled as "Entwurf equations". Entwurf equations still someway relied on physical consideration. Einstein's Entwurf theory, like Mie's theory, was based on world function represented by the components of the metric tensor field and their first derivatives. The world function L for the free Maxwell field is modelled on its Lagrangian first-order derivatives of the metric.

When Einstein acquainted the results of Hilbert's work at the end of October 1915 he demised his approach in Entwurf equations. His new field equations in paper of 25. November 1915 resp. in 1916 [12, V6, D25, D41] content the same number of gravitational and electromagnetic potentials as Hilbert's are derived fully from Hilbert's principles based on the variational principle of Hamiltonian of the world function of ether [65].

These Einstein new published equations were this time surprisingly generally-covariant field equations, but this general covariance was in full contradiction to his vigorous argumentation in previous years according to which generally-covariant field equations are physically unacceptable. Einstein offered no explanation what was wrong with this argumentation that he had published four times before [65].

All these theories mentioned above, unlike the theories of Hilbert and Einstein, are based on reality of physical experiments of four dimensional vectors (similar to the Lorentz procedure from 1895 paper) using mutual interactions of electric and magnetic forces of two objects (or sets of objects or manifolds) and gravity is taken as a consequence of these interactions. The covariance or invariance (in sense of relativity postulate) is in these theories mostly disclaimed or are not seen in the Einstein's or Hilbert vision that all reference frames are equivalents.

Hilbert (primarily mathematician) reverted this approach and for him electromagnetics is a consequence of gravity. The reasons as were mentioned above were exclusively mathematical. Hilbert and Einstein (in 1915 – 1916) exploited hypothetical coupling between gravitational and electric potentials. The form of the electromagnetic parts of their theories is a consequence of the structure of the gravitational part of the theory. The gravitational equations entail four mutually independent linear combinations of the electromagnetic equations and their first derivatives.

Einstein's gravity theories fall as usually into the pure mathematical structures. World function (ether) become a metric tensor determining Einstein's newly created 'physical reality' of the curvature of space and time without physical content or connection to physical reality.

It was shown above that all theories about gravity during the last four hundred years are about shrinkage of ether in space. Time or unit time in reality represent velocities imparted to bodies by the gradients and divergences of these shrinkages. These theories are not about space and time as it is forced by mass suggestion of power structures and academics on the general public through mass media and hyper production of demagogic documentaries.

The Identity instead of Equivalency of the Inertial and Gravitational Forces

Galilei and Newton found that the physical cause of the order of movement of celestial bodies is the existence of a gravitational field as a real physical substance, existing in the surroundings of each mass body. This gravitational field around mass bodies is inseparable from any single mass body. At the same time, with the discovery of the gravitational field around each of the existing masses, Newton (along with Galilei and other physicists) also discovered the existence of inertia as inseparably linked with every mass body.

The physical considerations of our predecessors in the past centuries, physicists who lived in inseparable connection with the great mass of our Earth, were thus also reflecting upon the main force that every man in his daily life come across - the gravity of our Earth. But unlike our predecessors, we can carry out experiments by our hands with objects of our daily usage in free space in the universe outside the reach of the great mass of our Earth. Or we can simulate this situation on International Space Station orbiting the Earth e.g. as we see it in television programs from ISS.

If we will manipulate in void space of universe with small object around us, let us say objects up to 1,000,000 kg, we would need to overcome inertial forces of this object by exerting up to 1,000,000 (10^6) N to cause these objects to move at added velocity of 1m/s.

The gravity force of these object are (or up to) less than one millionth 10^{-6} N on their surface and we will hardly reveal any gravity forces in these experiments. An explanation of gravity in this situation would have almost no meaning for us. Our principal physical interest would be concentrated in an explanation of inertial forces. If we explain these inertial forces, then the gravity forces are also to be explained as the supplementary negligible remnant of inertial forces. But, according to the GTR and today's mainstream physics, inertial forces as real forces do not exist and are just apparent fictitious pseudo forces and represent just a changed position of objects in space and time.

We have to ask what the validity of the GTR equivalence principle is and how do the inertial forces equal gravity forces in these situations? Where is, according to the GTR, the equivalency principle, the increase of gravitational forces as a consequence of a rise of gravity mass in these situations if we increase the acceleration of these objects and so the inertial mass is rising rapidly?

From their definition, inertial forces have zero value at the rest states of bodies. As a consequence of acceleration, inertial forces reveal and change from zero rest states value to value directly proportional to the size of acceleration. Contrarily, the gravitational forces never have zero value in the rest states of bodies. Gravitational forces definitely do not rise proportionally to inertial forces thus to rise of acceleration.

From these basic physical facts, it follows that gravitational forces can never be equivalent to inertial forces. As was disclosed in our previous paper [11], inertial forces as a change of zero rest states value (difference between zero and actual value) of bodies equal (are equivalent) to change of gravitational forces of bodies (difference between existing value at rest state and new value). Gravitational forces on the surface of atoms (protons of atoms) of mass bodies equal to internal spin momentum of these atoms. Gravitational forces are the result of superposition of change of tension of ether (gradients of field of ether) at the vicinity of atoms caused by the stress of internal spin momentum of atoms upon the ether.

So it is not the true that gravitational forces are equivalent to inertial forces, as states the principle of equivalence of inertial and gravitation forces of General theory of relativity.

The change in density of ether as a gradient of gravitational fields and gradient of inertial forces, both described by acceleration as change of velocity in unit time, became the basic physical principle upon which is built Classical mechanics.

Logically, the simplest physical conclusion that applies would lead Newton to determine the cause of inertial forces as forces of the resistance of the body against its own medium of the gravitational field.

The inertial force of a 1 kg spherical body is measured in its mass center (in the middle), but its own gravitational force is measured at a distance of 1 m from this center.

Inertial force of one kilogram of mass, however, is the enormous power 10^{12} times stronger compared to the gravitational force measured at a distance of 1 m from its center. Newton, in addition, didn't know the size of his own gravitational constant $G \approx 10^{-12} N$ (Cavendish's 1798). He also did not know the size of the depth of the structure of matter into atoms, 10^{-10} (Perin 1913) and protons 10^{-15} (Rutherford in 1920) that are the source and origin of the manifestation of all forces of mass bodies.

Today we can prove [11] that more than 99 percent of the forces of the gravitational field are located within the mass bodies and that these forces are, in fact, magnetic and electrical forces that keep the mass body as a compact object together. We can show [11] that the sum of the gravitational forces on the surface of atoms of the 1kg mass body is equal to its inertial force at the beginning of unit acceleration.

In this is a remarkable view of Newton that appears right in the first paragraph, when he defines mass states and says *"I have no regard in this place (* place of definition of quantity of matter*) to a medium, if any such there is, that freely pervades the interstices between the parts of bodies"* [1]. His 'Principia' closes thus: *"And now we might add something concerning a most subtle spirit which pervades and lies hid in all gross bodies; by the force and action of which spirit the particles of bodies mutually attract one another at near distances and cohere if contiguous; and electric bodies operate to greater distances repelling as well as attracting the neighboring corpuscles, and light is emitted, reflected, inflected, and heats bodies"*.

Today, we know, that photons (electromagnetic radio waves, X-rays, gamma rays), neutrinos, protons, electrons, alfa particles (and others) pervade freely through bodies and matters in an amount corresponding to penetrant attenuation coefficients of their mass densities.

Likewise, assignment of great inertial forces to the resistance of mediums (ether), however, would lead (as was supposed) to decelerating of bodies moving at a constant speed in this environment. This was supposed to be in conflict with the initial principle of physics since the time of Galileo and Newton where, without the influence of forces, bodies remain in rest or uniform rectilinear motion.

For the past hundred years it has been omitted that the absence of the resistance of ether in uniform linear motion was the main argument even at the condemnation of the ether at the time of interpretation of the M-M experiment. This irrefutable contradiction concerning the mechanical resistance of the ether in a uniform motion of bodies in free environments can however be removed after the discovery of the depth of the structuring of the mass and spin properties of all the particles of matter in the last hundred years. Since the nineteen thirties we found that all elementary particles are rotating spherical objects.

From the results of fluid dynamics and continuum mechanics we learned that, on the spherical symmetric rotating body moving at constant speed in an ideal fluid, only the same force from all directions perpendicular to the surface of this spherical body exists. The drag force on a rotating body, moving with constant velocity relative to the fluid, is zero. That means the rotating spherical object is not decelerated in uniform motion in an ideal fluid environment against the direction of its movement (already the evidence of d'Alembert that $\Delta = 0$ at d'Alembert's paradox in 1752).

The power of the resistance of the environment on a rotating body in a fluid environment in the direction of its movement is only manifested in the accelerating or decelerating of the body between the two speeds. The subsequent perpendicular pressure on the surface of the rotating spherical body moving at various constant velocity in a fluid environment is proportional to its speed of motion in that environment. As a result of the pressure changes of the surrounding environment on the spherical surface of the compressible rotational body, the change in the radius of its volume occurs.

This physical concept could uphold the conviction of Lorenz and Fitzgerald about contraction of dimensions of solid bodies as a consequence of different pressure of ether at different speeds of bodies within that ether.

H. A. Lorentz, in his 1904 paper [16] mentioned: *"The first example of this kind is Michelson's well known interference-experiment, the negative result of which has led FitzGerald and myself to the conclusion that the dimensions of solid bodies are slightly altered by their motion through the ether"*.

The actual physical reality of the relationship of inertial and gravitational forces is disguised in current physics by the damaging principle of the equivalence of inertial and gravitational forces in the GTR, in which the force of inertia of one body (the test body) is given equality with the gravitational force of another body (the central body). The gravitational force of the central body must be searched in relation to the inertial forces of the central body.

For Newton, the use of the law of action and reaction make no different in what kind of a force on the body (test) acts; for example the force of the impact of another body, dragging the body (e.g. lift) by a rope or the force of gravity [1, p. 84]. *"This law (action and reaction) also takes place in attractions, as will be done in the next scholium"*.

The forces in the law of action and reaction are the same, but the motion of bodies is not. The motion of bodies is inversely proportional to the mass of bodies, but only at the moving center of their mutual inertia frame. The independence of the nature of the forces acting on the body (the test body) is Newton's statement about the equality gravitational and inertial mass of the body. The reaction of mass of the body when exposed to the same gravitational or mechanical forces is the same $Fa/Fg = ma/mg$, so that gravity mass equals acceleration mass. As was shown above, there is also no contradiction in it.

But, from this statement, Einstein concluded that, consequently, in gravity fields gravitational forces on bodies are opposite and equal to inertial forces. So, as the sum of forces acting on bodies is zero and despite of that bodies move with g acceleration, so a curved nonmaterial space and time is caused which creates motion of bodies in gravitational fields. However, this reasoning is flawed, because the gravitational interaction of two bodies is always mutual. Gravitational and inertial forces of both bodies are simultaneously involved in the resulting mutual movement of such bodies.

Einstein GTR begins with Galileo's law that all bodies, independent of their mass, fall to the earth with the same g acceleration. Einstein takes this law as valid with the most accuracy *"which holds most accurately"* and based all GTR on it. But Galileo's law is valid only within the limited case when the mass of the falling body, as well as the gravity of the falling body can be neglected by the mass of the earth. If the mass of the falling body, equals the mass of the earth, then they fall mutually on each other with 2g acceleration.

In this case, according to Einstein, from the equivalency principle of gravitational and inertial mass, also arises that gravitational and inertial forces of earth as well as the falling body are balanced and so it is proven that no forces between them exist. So, according to Einstein, in this case when on earth a body falls with a mass that equals the mass of earth, their mutual action is caused by their two curvatures of non-material space and time.

If the mass (e.g. Sun) of the 'falling' body is far greater then mass of the earth, then this 'falling' body remains at the same place and the earth will fall to this body.

In Kepler - Newton celestial mechanics masses of objects determines not only ratios of forces of gravitational fields surrounding them but

4.

masses of celestial objects through their ratio
determines the speed and direction of mov
interaction.

The physical reality of inertial forces dete
direction of movement of each of two mutuall
any other inclusion of this reality in GTR fully a

The invalidity of Einstein's basic postu
demonstrated in a simple experiment in the ii
disk loadstone magnets facing each other with th
action of first magnet that we have in hand (to
is added the mass of our hand and our body),
distance) push the second not fixed magnet i
magnet moves despite, according to the law of
inertial force equals and is opposite to the push
field of the first magnet. Forces (contact pro
movement of the magnets will depend on the rati

But according to Einstein's equivalence princi
experiment, the second magnet should remain ir
the inertial force of second magnet has the
magnitude as the pushing magnetic force of the f
to GTR, curved space time is then caused, which
the second magnet.

In this experiment we can also place and let h
above the first magnet that we have in hand. W
magnet in our hand we canceled the space time c
we will feel in our hand the distance gravitational
the second magnet!

In both cases, the existing reality of the mediati
is so evident that it can be replaced in our mind by

If in this simple experiment of the interactio
magnets, these magnets face each other with the o
will attract each other. This situation will be exa

gravita
will be
magnet
no forc
claime

As
and g
mecha
confirr
validit

The
(e.g.
consid
GTR c
Ein
which
empiri
gravit

As
same
the fa
equiva
fall of

The
equiva
experi
attent
GTR.

93

ional attraction. If we keep one magnet in our hands the second
moving to us. If we release the one magnet from our hand
; will be moving one towards the other. The claim that there are
:s between these magnets moving one towards the other as it is
by Einstein for free fall in GTR is sheer nonsense.

·as documented above, the principle of equivalency of inertial
.avitational mass is a fixed constituent of Galilei-Newton
.ics and all experiments confirming this equivalency are
.ations of classical mechanics and not confirmations of the
of the General theory of relativity.

experiments confirming the equality of gravity and inertia mass
Cötvös experiment) are in GTR subsequently deceivingly
red as full evidence of the correctness of the basic principle of
·ncerning the equivalency of gravitational and inertial forces.
·tein's fully absurd physical logic is clearly seen at a conclusion
he derived from Eötvös experiment [12, V4, D16]. *"This*
·al law can also be expressed in the following way. In a
·tional field all bodies fall with the same acceleration"*.

·vas said above the fall of bodies in a gravitational field with the
·cceleration holds only within the limited case when the mass of
·ling body can be neglected by the mass of the earth. The
·ency of inertial and gravitational mass has nothing to do with the
·bodies in a gravitational field with the same acceleration.

sense of systematic and persistent routing and attention to the
·lence of gravitational and inertial mass in GTR at Eötvös
·nents as conclusive evidence of the validity of GTR distracts the
·n from the following Einstein's absurd primary constructions of

"*A freely movable body not subjected to external forces moves, **according to the special theory of relativity**, in a straight line and uniformly. This is also the case, according to the general theory of relativity, for a part of four-dimensional space.....in a gravitational field [12, V6, D30]. Because, for an observer in free-fall from the roof of a house, during the fall there exists no gravitational field. The observer, therefore, is justified in interpreting his state as being at rest [12, V7, D31, 1920].*"

And fully absurd logic continues in his primary argumentation of the equivalency principle: "*The experimental fact that the acceleration in free-fall is independent of the material, is powerful argument in favor of expending the postulate of (special) relativity to coordinate system moving non-uniformly relative to each other*".

In short, these constructions mean that Einstein's fantasy confirms that no force exists and so all physics is just geometry of mutual moving of inertial frames, regardless if it concerns of translational or of acceleration move. So, according to Einstein, the logic on the basis of his mathematical transformation from the first Galilei-Newton's laws that if there are no forces, bodies remain at rest or in uniform rectilinear motion, it follows that, in a gravitational field and at the acceleration generally, no forces exist. So that the physical states of bodies at first Newton law are the same as at the Newton third law of force F=ma.

So Einstein asserts that we feel no force if we hold heavy bodies in our hand in a gravity field on the Earth's surface, or that we feel no force of resistance as a consequence of acceleration if we throw an object on the Earth's surface or anywhere in the void space of the universe. This Einstein fantasy construction, which is the very first base of GTR, represents an ignorance of all experimental findings by all physicists throughout the whole history of mankind.

But at the end of this Einstein paper from 1920, as was detailed above, he claims that the metric facts can no longer be separated from the proper physical facts, so space and ether flow together into each other. So that gravity in GTR is caused by the force of change of density in ether. And, as was documented above, the contraction of rods in STR after 1920 is, according to Einstein, also caused by the pressure of ether.

The equality of the gravitational and inertial mass of one body and the movement of bodies in their mutual action are two entirely different items. The physical reality is that, although the forces are the same in earth gravity, each body falls to earth by neglecting the mass of body to the mass of earth.

The mass of a body defined as the resistant inertial force on the basis of Newton law $F = ma$ is independent of its position in the universe. The mass of the body defined in gravity field $F = mg$ is dependent on the strength of the gravitational field and its position in this field.

So by setting a mass of the small body in the gravity field, we do not check the mass of the body bad we check the strength of gravity field. The relation $E = mc^2$ for amount of energy in matter in relativity is taken as valid everywhere in the Universe, but amount of mass in this relation is taken as force of weight on Earth in kilograms so on the Moon where this mass represents five-fold smaller force the amount of energy in matter will be five-fold less. By weighing different bodies in the same gravitational field just their relative amount of masses as well as their relative internal energies are set.

The principle of equivalence in the GTR confuses and mixes Newton's statement concerning the equality of gravitational and inertial mass with a false embrace of the law of action and reaction. The law of action and reaction is held in the GTR as the same time and same place equality and reverse orientation of the forces exerted by actions and reactions. Unfortunately, this flaw occurs countless times in current physics (although never mentioned by Newton). But this would lead to a standstill of the entire universe.

The crash or any force of action between any two bodies would have had to stop the movement of these bodies. The body in the gravitational field would not have moved, since the force of gravity and inertia are opposite and balanced.

In the basic physical thought experiment of GTR on equality gravity and inertia, the man standing in a stationary elevator in a gravity field is pressed to the floor of the elevator by the same force as the man standing in an elevator pulled by rope, with acceleration equal to the acceleration of the gravity field. But, at the same time, in GTR it is claimed that during acceleration at fall of bodies in gravity there are allegedly no forces (we feel no forces) and gravity can also be compared with the acceleration resistant force of these bodies in void space.

As was detailed above, Einstein at the same time in this situation also claims that *"the inertial mass of a body falling in gravity is increasing and this must equal to an increase in its gravity mass"*. Simultaneously, he claims that the man standing in an elevator pulled by rope in void space in no case can distinguish if he is not in a static gravity field. But from the claim in previous sentence for this distinguishing, as was also detailed above, evidently just sensitive enough scales are needed.

What then is the content of the statement of the equality of inertial and gravitational forces in so called Einstein's equivalence principle? Is it the stationary situation in a gravity field when the force equals to the non-stationary situation at the same acceleration force in void space? But this equality is held just in the very first infinitesimal moment and after that this equality is in fact, from the first moment, invalid due to increasing of inertial mass at acceleration in void space. Or is it the situation at falling in a gravity field where the gravity force is allegedly zero and so no equality with acceleration force in void space can be done?

On this schizoid basis of GTR is constructed allegedly the greatest intellectual achievement in the history of mankind!

The only thing that is clear from these statements, as well as from plenty of other contradicting statements across the entire STR and GTR, is that Einstein did not understand the basic physical principles. He was not sufficiently familiar with the basic formal logic and the foundations of classical mechanics of Galileo, Kepler and Newton.

The elevator thought experiment is a misleading asymmetric description of physical reality. The situation in the stationary elevator cannot be made equal to the non-stationary elevator pulled by a rope. The force of gravity pulls the elevator (or today rather pushes), but for all parts of the atoms of the lift and all parts of the man standing in it.

I submit that the situation of the stationary elevator, in a shaft at a floor, blocked in the gravity field is symmetrical with the elevator pulled in the free space of the universe by pulling rope with the thousands of invisible glass fibers, that pull at the same time for all the atoms of the elevator and all parts of man, and the free movement of the elevator is prevented by a block against the rope mounting.

A man standing in the stationary elevator in the gravity field would feel pressure on the soles of his feet as the pressure of his own body and no pressure on the top of his head. In the stationary elevator in the free space pulled by a rope with thousands of invisible fibers, the man standing on his head, would feel the pressure on the top of his head and no pressure on the soles of his feet.

In the case of deletion of blocks in both cases, the person would feel no pressure and would feel free-fall or free-acceleration. The biggest blunder of the GTR is the claim that at the free-fall of a body in gravity (equivalent case to "free acceleration") no forces exist and so the movement in gravity (just as at case of "free acceleration") has to be assigned to the curvature of the space and time.

The gravitational field, the forces of that gravitational field effect on the level of the atomic particles of objects. For as long as Einstein's roofer continues to stand on his scaffold contact pad, the gravitational field affects the atomic particles of his body. These push, one against the other, only due to the resistance of the scaffold, which is the reason the roofer feels the gravity as the pressure of his own body through his pressure receptors. The reason why Einstein's 'falling roofer' fails to feel the gravity, is that the roofer has no receptors on an atomic level through which he would feel this gravitational force. Equally, upon his acceleration in the free space of the universe, Einstein's roofer feels acceleration through the pressure receptors as the mutual pressure of his atomic particles against the contact pad causing his acceleration.

But as is detailed in chapter 7, we do not reject the existence of air because we do not feel the pressure of 15000 kg of air on the surface of our body. We do not feel, as well, the reality of a billion times a billion photons of electromagnetic radiation, neutrinos, relict radiation, protons and other particles passing through our bodies every second.

Finally, today we now know that to change the height of the orbit of a satellite circulating around the Earth we must turn on the reactive engines acting on the satellite by force in the direction or in the opposite direction to the gravitational forces (not against nonmaterial space-time) and add or subtract energy to the satellite (as cumulative force) according to the desired size of the orbit height changes of the satellite.

GTR

Since 1925 (Hubble's discovery of galaxies, 1926 Lindblad's discovery of rotation of galaxies), we understand that all the stars and constellations that we see in the night sky with the naked eye or binoculars (numbering about 2,000 stars) are just a small part of our nearest surrounding universe in a sphere with a diameter of 2 to 10 thousand light years in our Milky Way Galaxy, which has a diameter around 100 to 200 thousand light years. These naked eye visible stars in this sphere rotate along with the Earth and the Sun around the center of our galaxy. The ancient Greeks, Newton and Einstein (at the time of 1905, 1915) considered these with naked eye visible stars within this sphere with radius from 2 to 10 thousand light years in the sky (together with a few nebulas and the belt of the Milky Way nebula), as the entire universe.

Newton, after the formulation of the general principle of gravity of all masses, could not avoid the question of why the universe had not gravitationally collapsed. Newton's reply to this question is based on the knowledge of his predecessors, that for the previous 3,000 years the universe appears to have been stable and, secondly, on the knowledge of his own discovery that in the solar system the gravitational force of the Sun acting on the planet is compensated by the centrifugal inertial force of planets orbiting the Sun.

Therefore, Newton was convinced [1, p. 514] that the gravitational forces acting on the stars in the sky do not route to the Earth or Sun, but to their own force center on the particular orbits of these stars. Since 1925, we understood that Einstein's description of the image of the universe in general relativity theory for the universe known to Einstein (a sphere within a radius of 2 to 10 thousand light years) is mistaken. This is because the universe known to Einstein as a whole rotates around the center of the Milky Way Galaxy and the mutual gravitational interaction of stars within this sphere play no role in the physical image of this universe.

The discovery of whirling galaxies meant the end of GTR. Newton's far-sighted belief has proven correct with the minor change that all the stars of the universe known to Newton (so as to Einstein) rotate around the common power Center of the Milky Way Galaxy.

After this proof of the blunder to use GTR for stars in our Galaxy (a universe known in 1915) it was claimed that GTR is but valid for newly discovered universe of billions of galaxies. It is very alarming that these allegations of current physics, supporting this theory as applicable continue to be put forward. However, again the Big Bang theory of physicist and priest Lemaitre (1927) is put forward as truth.

Physics during the last 100 years has shown that the basic manifestation of the mass of the observable universe around us is its rotary and curl movements. Fields of alleged quark-gluon particles curl inside the proton and neutron. The proton and neutron most probably rotate as fields at shells in atomic nuclei, electrons rotate as fields at shells around atomic nuclei. The Sun and planets rotate individually; the planets around the Sun, the Solar System and other stars rotate around the center of the Milky Way Galaxy.

With full physical conviction we must then assume mutual rotary movements, even for galaxies and their higher grouping. However, from the measurements of the observed redshift of spectra of a Galaxy we are unable to determine its direction of possible movement in space.

Even with the evidence of the last twenty years that Galaxies form filaments in web-like super cluster complexes, that they move at curved paths along these filaments and that all hundreds of Galaxies closest to our Milky Way Galaxy are shrinking in direction to the Great Attractor we are not yet able to get the academic community to a clear rejection of the Big bang theory forced daily upon them under the supervision of power structures and public mass media.

The initial natural idea by non-physicists, would have seen the curvature of light in the gravitational fields of the celestial mass body. They would consider that, in the surroundings of this mass body is something that curves the trajectory of light, rather than the idea that there is nothing that causes this curvature (proven by the ordinary experience of curved glass or on the passage between two various matter densities or within gradient of fluids).

There is no logical reason why from a physical point of view held by all physicists in the late 19th century on this matter (the gravity as a gradient of the real physical ethereal substance) the light path should not be curved in the gradient of this substance of the gravitational field.

It was not necessary to carry out large expeditions like Edington's in 1919 for the purpose of observing a light bend near the Sun at the eclipse of the Sun. According to the fundamental experimental knowledge of optics since Newton, we know that light curves when it passes closely around the edge of any object that we have at hand.

To consider, however, the phenomenon of the curvature of light in gravitational fields as evidence of the validity of a physical claim that gravity is the curvature of the nonmaterial space and time is clear Dadaism. It can best be captured by the words of Nikola Tesla (1856-1943) in the New York Herald Tribune, 11. Sept 1932[17],

"I hold that space cannot be curved, for the simple reason that it can have no properties. Of properties we can only speak when dealing with matter filling the space. To say that in the presence of large bodies space becomes curved is equivalent to stating that something can act upon nothing. I, for one, refuse to subscribe to such a view".

Tesla in his works claimed that Einstein's relativity, which discards the ether, is entirely wrong and he proved that no vacuum (void space) exists. He asserts that all attempts to explain the workings of the universe without recognizing the existence of ether and the indispensable function it plays in phenomena are futile. He asserts that there is no energy in matter other than that received from the environment.

Special relativity as a mathematical construction is without any physical contemplation firmly rooted in the purported zero result of the experimental investigation of the speed of light, carried out by Michelson- Morley. General relativity emerged from Einstein's purely theoretical and somewhat misguided speculations about a possible relativity of acceleration.

Gravity, electricity and magnetism were explored independently in physics and are also currently presented in physics as independent parts of the science. Gravity in physics is referred to as (and often as the only one) universal power and universal attribute of matter. However, after the last hundred years of exploring molecules, atoms, protons, electrons, and other elementary particles, we know that each atom or molecule, each proton or electron, and other elementary particles of matter exhibit a magnetic field in its vicinity and in their inside or around the electric field.

If we look into tables and catalogues (CODATA, Wikipedia or others) showing the basic properties of elementary particles and atoms, we find that for every particle or atom is given one or both values of the force of the electric or the magnetic field. Even the neutron has a magnetic field approximately equal to the magnetic field of a proton!

However, for these elementary particles in these tables we find that no value of the size of the gravitational forces fall to the size of the mass of these particles, even when the size of these gravitational forces on the surface of the particles or atoms equals at least their magnetic force [11].

We can prove [11] that more than 99 percent of the forces of the gravitational field are located within the mass bodies and that these forces are in fact magnetic and electrical forces that keep the mass body together as a compact object. These same forces hold together electrons, protons and neutrons inside atoms as well as hold fictitious gluons and quarks inside protons and neutrons. The gravitational field surrounding the mass bodies is much smaller than a one percent remnant of these forces.

The explanation of gravitational fields lies in the extension of Van der Waals experimental study (1910 Nobel Prize) on the existence of mutual attractive forces between the molecules and atoms of substances emerging as an averaging remnant (magnetic field) at their random thermal rotating (spinning) movement of their thus rotating dipole and multipole electrostatic fields. No other forces than that ones that appear around the elementary building units of matter, atoms and molecules, can appear around great mass objects which are compounded of the great quantity of these elementary building units.

We can consider, with great conviction, that the gravitational force is not a universal fundamental attribute of matter, and that the force of gravity as the individual fundamental power of atoms and particles does not exist. We can consider that gravitational fields and gravitational forces surrounding great mass objects represent the sum of a huge number (1kg steel ball 10^{26} atoms) of disordered magnetic fields of the atoms and elementary particles from which these mass objects are composed.

We can consider that the gravitational field of the Earth is the dominant vertical component of the sum of a huge number of the disordered magnetic fields of atoms and elementary particles of the Earth. The magnetic field of the Earth is a manifestation of the asymmetry of the layout of the dominant vertical component of the magnetic fields of atoms of the earth caused by the spherically asymmetric ellipsoidal shape of the Earth. For the measurement of the magnetic field of the Earth, e.g. by a compass, we have to eliminate this main dominant component of the field and spar the needle of the compass in the middle (in its center of gravity).

Subsequently, under the so-called gravitational waves of large gravity mass bodies established in GTR, it is necessary to consider the broad spectrum of the disorderly flow of so–called thermal electromagnetic radiation of the individual atoms and elementary particles from which these large gravity mass objects are composed. Maybe relic radiation (cosmic microwave background) is just a manifestation of this way in which the existence of so-called gravitational waves are presented. They can be seen as the broad spectrum of the disorderly flow of very low electromagnetic radiation frequencies of the individual atoms of large mass bodies rotating in solar system, the Milky Way and other Galaxies. Analogous to cyclotron radiation, they are exposed to acceleration components along an orbital path.

GTR at introducing the magic object called the Black Hole, claims that no escape (light or any bodies) is possible beyond the boundary of the region called the 'event horizon' and that Black Holes can be identified upon the basis of their gravitational interaction in otherwise boundless distances. The Black Holes are allegedly placed in the center of galaxies. Black Holes by massive gravity fields keep the rest of the galaxy together.

The first simple and logical question then is how the gravity forces themselves or the gravity field of the Black Hole escapes within boundless distance beyond the event horizon? This was never raised in GTR and so much the more never answered. GTR, on the other hand, at the same time also claims that Black Holes can produce gravitational waves that transport energy to boundless distances as gravitational radiation, a form of radiant energy similar to electromagnetic radiation. But electromagnetic radiation and light are the same.

Later, the fantasy about particle of gravity called a graviton was added to GTR. The substance of a graviton has to be the curvature of space and time as it is a particle of the curvature of the space-time field. Maybe if to a graviton is added another fantastic feature, that it can move with velocity higher than light or infinite velocity, it can solve the trouble with the range of gravity beyond the event horizon.

Einstein in GTR transformed Riemann's assertion, that space acquires a definite form only through the advent of the material content filling it and determining its metric relations into a demagogic physical and philosophical phantasm of the mystery of the curvature of non-material notion of space and time.

However, these curvatures of space and time cannot be directly experimentally measured.

This was a continuation of the mystery of time dilation in STR, where the mutual velocity of body and light is always constant, regardless of the direction of movement of the body towards the light in the equation $v + c = v - c = c$. The difference in velocities is the difference in the traveled paths. Path divided by time is velocity. So to prove the validity of this equation, the mystery of time dilation is introduced. But from time dilation follows the deceleration of velocities. In this equation the reference frame is also not defined.

In relativity, using the multiple of speed and time, we are not able to measure the traveled paths because from an increase in velocity follows dilation of time and length contraction and so also follows the deceleration of velocity itself. We find ourselves in a typical mysterious Einstein pathological circle, where no basic unit of quantity and no basic reference frame is definitively defined. In relativity, all basic units of quantities and their relevant reference frames (so all physical law) is changing according to their ratio to the velocity of light. But no reference frame exists for the velocity of light.

But physics is a fully comparative science. No single absolute numerical value exists in physics. Physical constants (the basic physical unit of quantities) are firm calibration values of physical law for comparing the development of observed phenomena. Without fixed calibration values of basic physical units of quantities and fixed reference frames to which these calibration value are related, we can't discuss physics.

Albert Einstein, on his 70th birthday, in a letter to Maurice Solovine, 28 March 1949 [18, p. 328]-"*You imagine that I look back on my life's work with calm satisfaction. But from nearby it looks quite different. There is not a single concept of which I am convinced that it will stand firm, and I feel uncertain whether I am in general on the right track*".

B. Riemann in his work [19, p. 11] says *"in a discrete manifoldness (existence of particles in all neighborhoods), the ground of its metric relations is given in the notion of it, while in a continuous manifoldness, this ground must come from outside. Either therefore the reality which underlies space must form a discrete manifoldness, or we must seek the ground of its metric relations outside it, in binding forces which act upon it"*.

The conclusions of Riemann imply that, we can get rid of infinite space just in case the metric space is filled with discrete particles and it must be added that these discrete particles must be at rest.

Riemann's conclusion should today compete with the physical reality of the production of discrete particles as curl compression of the continuous ether, so we must seek the ground of metric relation of space filled with such discrete particles also outside of it.

Today we know that particles move at huge speeds within the expected continuous manifoldness of space. Consequently Riemann's idea of the construction of a metric of space grounded on the property of discrete particles inhered in mutual neighborhoods is also rather illusory. Again, it would be necessary to construct a metric space based on the pressure of the media at a given point and the total pressure of the space with an opposite pressure outside of this space.

In the deprivation of infinity we also seek a reason why the power structures in physics support the theory riding off the continuous ether;

special and general relativity, quantum mechanics, force fields theories using force-mediating particles, the standard model of quark theory, the Big Bang and the Higgs boson theories. These theories are merely based on the existence of rigid particles that, together with the theory of curved space-time, provide the requested finiteness of the dreamed of universe in which when we move in any direction, we will always move on a curved path within this space.

This unproven speculation has a basis in the discovery of the roundness of the Earth, while till then some assumed that the Earth was flat and a sea cruise ended on the horizon with a subsequent fall to hell. They were even afraid of many sailors on the Santa María during Christopher Columbus's revelatory voyage in 1492 that discovered America.

Frank, intelligent, honest and fair physicists must clearly tell the public and physical community that an understanding of infinity is beyond the current ability of the human spirit and of today's civilization. An experimental possibility for the inspection of the infinity of the macro world is always as inaccessible as infinity itself. It is not in our hands, because the ratio of any range of telescope to infinity is always zero.

In contrast to the inspection of the infinity of the macro world, the situation for the inspection of infinity or finiteness of physical zero (existence or non-existence of smallest particles or fluid quanta of the physical world) of the micro world is not so gloomy and is in our hands. Understanding the essence of the micro world of physical fields (ether) here, directly under our hands, can support our understanding of infinity and of the macro world of the universe.

In contrast to STR, GTR is based on Einstein's stunt declaration that the velocity of light is not constant any more as can be read e.g. in his 1914 paper [12, V4, D16] – "*This bending of light rays implies that the velocity of light is not constant, but depends, instead, on the location. This forces us to generalize the theory of space and time, known as the theory of relativity, since the later was based on the assumption of the constancy of the velocity of light."*

The GTR field equations incorporated in experimentally measured Newton's gravity constant G included in terms of the proportionality constant (Einstein's constant or Einstein's gravitational constant) $\kappa = 8\pi G / c^2$. Constant G presents the experimentally measured fact that two 1 kg mass objects gravitate to each other at the distance 1m with force G. This experimental fact means that between these two masses or all other rigid matter in the universe we measure the force which causes all mass to move closer and couple to each other (if there would be no counteracting accelerating centrifugal inertial forces).

The cosmological constant was, again after 1990, added into Field equations although it was definitely discarded by Einstein after 1920 as the biggest blunder of his life. This time after 1990 the cosmological constant presents the curious force which stems from dark energy. According to the contemporary mainstream cosmology claim, across the entire universe, in each of its individual points, exist forces counteracting and overmatching the gravity force that expands the universe (which will in future allegedly even cause a rip to pieces of all atoms of the universe, of the Earth).

But in this case no value of Newton's universal gravitational constant G would be measured for 200 years in physics. Contrarily, not the gravity constant G (the force which is the causes of moving of matter together) but the universal repulsion constant would have been measured (the force which is the cause of moving mass apart).

The logical consequences of the above mainstream physics claims about the cosmological constant are that the proportionality constant (Einstein's constant or Einstein's gravitational constant) $\kappa = 8\pi G / c^2$ do not exist and Einstein's field equations must be scrapped.

Also fully controversial is Einstein's ultimate claim of GTR from 1915 that no forces in gravity fields exist but the curvature of space and time is the cause of gravity. This contradicts the fact that all calculation of the curvature of the space time in Einstein's field equation $G_{\mu\nu} = \kappa T_{\mu\nu}$ is calculated from the experimentally measured gravitational forces acting between two 1 kg mass bodies at the distance 1m from Newton's constant G incorporated in the term $\kappa = 8\pi G / c^2$.

Likewise it is necessary to stress that the constant κ through the constant G is the single one physical reality in the 'Einstein' Field equations. Without this constant, the Field equations are useless and this constant is in fact not a constant and depends on the amount of mater in the space of the Universe. As the amount of mass increases the constant κ increases but the velocity of light c decreases. So most likely not ratio but some form of the product of G and c could be the space constant. The constant $\kappa = 8\pi G / c^2$, submitted by Einstein for construction of this constant, he moreover never provided the physical reasoning and is just his guess, wish and sheer fantasy.

Above all this, Einstein's constant $\kappa = 8\pi G / c^2$ was changed in later decades by mainstream physics to $\kappa = 8\pi G / c^4$, so Einstein's mistake of 16 orders was corrected! Moreover, mainstream physics at this change of Einstein's constant is not worried about the so called dimensional analyses of this procedure.

STR

An entire generation of hundreds of physicists of classical mechanics and electricity and magnetism, at least from Newton until 1905, when most of them after decades personally carried out direct experimental observations of the physical world around them, came to the claim of the existence of force fields as real physical substances around the physical body. This claim did not appear after two years of speculation about a single Michelson-Morley's (M-M) experiment behind a table in an office.

It is needful to point out that the reason for the implementation of the Michelson-Morley's experiment as was reported in Lorentz 1895 [63] work was to bring a decision between the opinion of Fresnel about the ether as a static medium, through which is sailing the Earth and the opinion of Stokes about the ether, which is partly or completely dragged by the motion of the Earth. In no case did this experiment intend, had aspiration as well as it was not in the power of this experiment to judge the question between the existence or absence of the ether as it has been a hundred years falsely perverted by STR exponents. So, according to prime two suppositions the null or small results of this experiment confirmed that the ether is fully or partly dragged by the Earth. The Stokes opinion about ether as the medium dragged with the Earth, so as the static medium in Earth frame was also the Maxwell creed expressed by Maxwell, "The time required by a light ray to travel forth and back between two points must change, as soon as these points are subject to a common displacement".

A century-long debate on explaining the M-M experiment is today completely useless. Today, thousands measurements of the velocities of automobiles by police radars executed daily around the world in which the difference between the velocity of automobiles and light ray in direction forth and back is measured confirm the Stokes opinion about dragging ether with the Earth and opinion of Maxwell about the medium in which the light is propagated. These daily thousands radar measurements fully reject Einstein's Special relativity fantasy.

Moreover, concerning Michelson-Morley's experiment, we now know (the discovery of composite rotary motion of the earth in a surrounding space, around the Sun about 30 km/s, around the center of the Galaxy about 220km/s, to the Group of galaxies about 700-1000 km/s) that the basic physical assumptions for the explanation of this experiment were wrong.

The assumption of the Earth's rotation around the sun at the rate of 30km/s, as the only motion of the Earth in space, is not valid. Also invalid, is the assumption of the rectilinear of the motion of one arm and the rectilinear of motion of the second arm of an interferometer toward the surrounding space, that is, the ether. So it is invalid that relation $\Delta t = 2L / c\sqrt{1-v^2/c^2}$ represents the time difference of flight through light in the perpendicular arms of M-M interferometer. This difference summarily represents the different mutual velocities of the light and interferometer, their different directions and different travel distances of the light inside the interferometer from those that were supposed in original presumptions at explaining the M-M experiment results.

It is necessary to stress that Michelson-Morley's experiment in 1887 (and all others later) never provided zero results, but results which were considered by Michelson and Morley to be negligibly small in light of the early physical assumptions of Fresnel static ether theory in their experiment.

This means that STR never became valid, because its validity requires zero results. This is clear also according to affirmation by Einstein himself as detailed in the chapter 8. "*Should the positive result be confirmed, then the special theory of relativity and with it the general theory of relativity, would be invalid*". The negligibly small results of M-M experiment Einstein enunciated as the zero result and he based his STR on this never measured zero result.

As the zero result of the M-M experiment is presented in all textbooks of physics till today. The fact that the result obtained by Michelson and Morley in 1887 was not negligibly small was very fully set forth by Professor Hicks of University College Sheffield in 1902, in his important theoretical examination of the original experiment (detailed in chapter 8.).

It is necessary to state that all the experiments similar to the M-M experiment (yet since time of Fizeau, 1848) in which medium of transmission of light propagation was under the 'control' (interferometers embedded or filled with gaseous or liquid medium (e.g. Mach–Zehnder interferometer) confirmed the expected result of time difference of flight through light in two perpendicular directions. In addition, the Sagnac (1913) experiment with a rotating Interferometer in a vacuum also provided the expected results.

In all physics textbooks, the illustrative explanation of the M-M experiment is presented by situations which show the travel difference time when one boat or swimmer swims a distance across the river perpendicular to the constant stream of flow from one bank of the river to the other, and at the same time a second swimmer swims from the same starting point the same distance along the banks of the river downstream and upstream of the river.

In the case of mutual rotation of the interferometer and the fluid in its surrounding environment, we can present this situation as only the rotation of a fluid. For the illustrative explanation, we can present a situation with a swimmer swimming at a constant speed in a circular pool, rectilinearly from the center of the pool to the edge of the pool and back while the stream of water in this pool, rotates at a constant speed around the center. A swimmer can swim rectilinearly from the center to the edge of the pool and back in two perpendicular directions or in any two directions and his swimming times will always be the same. That means, in the case of mutual rotation of the interferometer and the ether time of passage of light in perpendicular arms of the interferometer will be always the same in any rotation.

It is evident that the explanation of the M-M experiment was based on false assumptions and expected the wrong conclusions. Pertinent null result of M-M experiment means null result and from this null result cannot be concluded nonexistence of the ether. We can be convinced that the pertinent null result of the M-M experiment represents proof of the rotating mutual movement of the Interferometer and ether in its surroundings together with dragging the ether by Earth.

For the purported rejecting of the ether in 1905 it would be necessary to clearly deal with the other of its manifestations such as inertial forces, gravitational forces or, later after 1930, the creation of pairs of particles and antiparticles from and into electromagnetic waves. This is what has been done in 1915 by Einstein in his space-time mystery of GTR rejecting accelerations and forces. This was later also the intention of mystery of Quarks and Higgs Boson theories.

But as was shown above, after 1920 Einstein rejected his STR and GTR theories of relativity from 1905 and 1915 and fully returned to the undoubted existence of ether.

Another fundamental consideration about the outcome of the M-M experiment is the view of Fressnel (1818) or Stokes (1844) that the ether is partially or completely dragged by Earth and thus shares its motion at Earth's surface which gets a factual physical image on the basis of the results of this work.

These results show that all smallest elementary particles are spin products of curl compression of ether and their existence inevitably brings the existence of force fields (electric, magnetic, gravity) in its surrounding (as well as for their gathering in great mass bodies) as the gradient of ether otherwise uniformly filling space in other parts of the universe. This gradient, firmly fixed with any bodies, is moving through space along with great mass objects as well as the smallest elementary particles. As a result, photons of light as spin products of the electromagnetic curl compression of the ether are slowing or speeding when moving in this moving gradient.

A hundred years after the inception of the special theory of relativity, we lived to see the speed of material objects of protons equal to almost the speed of light. Protons accelerated in the LHC tunnel at CERN reach 99.9999991% of light speed; almost 1c. Two opposite direction beams of protons flying against each other, each at 99.9999991% of the speed of light, with mutual speed $1c + 1c = 2c$ collide in a tunnel. But STR in the equation for composition of velocities $u = (u'+v)/\left(1+(u'v)/c^2\right) = (c+c)/\left(1+(c \cdot c)/c^2\right) = c$ claims that this mutual speed is in fact $1c + 1c = 1c$.

So although we measure the mutual velocity of protons in the LHC tunnel as 2c, Einstein's relativity says, that it is not the true, and allegedly his equation based on his fantasy of time in Relativity evidences, that only correct mutual velocity in this case is 1c. This is the purposeful destruction of physics as a science, destruction of common sense and dogmatism of the highest degree forced across the whole world on mankind by power structures as the indisputable gospel truth of putative genius.

According to STR the velocity of the opposite beam in the LHC tunnel measured from reference frame either of this two beams is zero

$$u' = (u-v)/\left(1-uv/c^2\right) = (c-c)/\left(1-cc/c^2\right) = 0 \ .$$

In the STR beam of protons from which we do measurement stands or, better said, its own velocity is in STR eliminated, allegedly because of his own time dilated to infinity (exactly 55555556 times) and its own length contracted to zero (55555556 times).

For mutual velocity of two photons Einstein's 'genial' new physical law of composition of velocities declared as one of the greatest achievement of STR gives always the velocity c.

Einstein's law of contraction of length at the velocity of light gives zero length unit $l = l_o \sqrt{1-v^2/c^2} = 0$ and his law for dilation of time at the velocity of light $t = t_o / \sqrt{1-v^2/c^2} = \infty$ gives infinite time unit so the light moves always on zero length at infinite time. The result is that the velocity of light is always zero $v = c = l / t = 0 / \infty = 0$. From these Einstein's new laws of physics follow that light, all photons according to STR always stand.

Today these equations are clear evidence of a distortion of the physical reality in STR but in spite of that these equations are taught as reality for a hundred years even at secondary schools all over the world.

The consequences of these matters of fact should have resulted at immediate removal or at least the suspension of relativity in physics.

Time dilatation is the most serious forgery of STR. The basic physical relations established by Einstein are already in full contradiction in the issue of time dilation. In relation to the energy of the photon $hv = mc^2$ (Planck's idea), the frequency with increasing energy increases $v = m_o c^2 / h\sqrt{1-v^2/c^2}$ so time unit (one tick) shortens, but in STR relation for time dilation $t = t_o / \sqrt{1-v^2/c^2}$ time unit dilate with increasing energy.

Similarly, according GTR with increasing gravity towards the central source of gravity, so with increasing energy, time unit dilates. A clock moved near a source of a gravitational field runs more slowly, as its frequency is lower (as the force is increasing this contradicts to all physical contemplation of physical reality).

This claim we can read in Einstein's 1916 GTR paper - *"Thus the clock goes more slowly if set up in the neighborhood of ponderable masses. From this follows that the spectral lines of light reaching us from the surface of large stars must appear displaced towards the red end of the spectrum".* [12, V6, D30].

But these claims contradict the claim in 1907 - 'On the relativity principle and the conclusions drawn from it' [12, V2, D47, p. 307] *"..we may say that the process occurring in the clock, and more generally , any physical process, proceeds faster the greater the gravitational potential at the position of the process taking place".*

This conclusion from 1907 would be the natural physical conclusion of each serious physicist contrary to opposite claims from 1916.

So in 1907, according to Einstein, around large stars no physical gravity fields (they are rejected in STR and GTR) which removes energy (so the frequency) of light passing through them exist. Light emitted from large stars is redshifted because the clock goes more slowly. So Einstein gives direct proportionality between slowing down the clock and slowing down the light frequencies. This was detailed in chapter 3.

Einstein, on top of this, in 1911 also simultaneously claimed for this situation that in a gravity field the frequency of light is everywhere the same but just the clock by which we measure the frequency runs slower.

But after 1916 Einstein makes an opposite claim in his theories (intentional destruction of physical reality in favor of metaphysics of time) based purely on his physical fantasies called 'thought experiments' by him. Einstein in 1916 brings his new discovery (clock goes more slowly and so redshift *must appear*) forthwith as the first experimental measurements of gravitational redshift then appeared. Each serious physicist would, in the first place, conclude that the gravitational redshift is direct evidence of the existence of a gradient of real material physical fields around large ponderable masses and is a consequence of energy loss in the gradient of these fields.

Today introductory reasoning at Wikipedia   en.wikipedia.org/wiki/ Gravitational _redshift is fully contradicting and metaphysical where fraudulent is mainly - *"Redshift is a direct result of gravitational time dilation. As frequency is the inverse of time, specifically, time required for completing one wave oscillation, frequency of the electromagnetic radiation is reduced in an area of higher gravitational potential"*.

Einstein, for the purpose of his plan from 1905 to remove the ether of physics, did not hesitate to say in 1916 that the cause of gravitational redshift lays in slowing down the clock, which is the complete opposite physical claim than he made in 1907

In today's mainstream GTR, according to gravitational redshift, this physical law is again completely opposite to Einstein's GTR in 1916; the frequency of photons as moved near a source of gravitational field is higher (clock goes faster) and as moves away is lower.

Relativity introduced the claim that the mutual velocity of bodies moving at any speed and any direction relative to the movement of the light always equals to the speed of light. Einstein said that Maxwell's equations are valid in every inertial system only if the propagation of light in each inertial system is the same c. So each inertial system must maintain the difference of velocity of inertial frames and velocity of light is again the velocity of light v-c=v+c=c or v/c=c.

Today thousands measurements of velocities of automobiles by police radars executed daily around the world reject this Einstein assertion.

However, this claim also completely excludes any consideration of the possibility of the existence of waves in medium.

This claim completely excludes any construction of Maxwell equations or Lorentz force, because this construction requires a ratio of velocity of the source or of the receiver to the velocity of light v/c. The waves in a medium originate as changes of density of the medium caused by ratio of velocity of the source of these changes against the constant propagation velocity in this medium, which is also the carrying medium of these waves.

This claim completely excludes also the phenomenon known as the Doppler shift, as well as the Liénard–Wiechert retarded and advanced potentials. The essence of the Doppler shift is in the varying number of waves of media impact on the receiver, depending on the varying speed and direction of motion of the receiver relative to these waves.

In relativity, albeit from its first principle of the same mutual velocity of receiver and light waves, for an explanation of the Doppler effect the receiver can suddenly move between the two fronts of waves (moving against the receiver always with constant speed) with various speed v + c or v - c. Since, however, must pay c + v = c - v = c this perverse code reincarnates into the mystery of time dilation. As is shown below, time dilation is identical with the change of speed and so for time considerations sick code $1/t_c + 1/t_v = 1/t_c - 1/t_v = 1/t_c$ also is valid.

Moreover, relativity brings various declarations associated with the phenomenon of Doppler shift, which mainly includes a debate on the relativity of the red shift in conjunction with the expansion of the universe.

Even the starting point of Einstein's kinematic meditation about Galilean transformation $x' = x - vt$ cannot be connected with physical reality of Dynamics so with energy or mass (STR never included the definition of mass). As was detailed in chapter 3, Einstein in 1911 even rejected this approach *"If we had based our considerations on the Galilei transformation we should not have obtained a contraction of the rod as a consequence of its motion"*.

While until the year 1905, physicists investigated how bodies ergo inertial frames reach the desired velocities, Einstein in STR ignored all these previous investigations. In STR the reference frames have simply desired velocity and Einstein, without any physical reasoning, faked up his own physical law for transition between two frames with two different velocities.

In Dynamics linear change of velocity v (transition between two inertial frames in STR or two wavelengths or frequencies in QM) needs a quadratic change in distances x^2 (or wavelengths $1/\lambda^2$ or frequencies ν^2 in QM) which leads to a quadratic rise in energy proportional to v^2 . That is why classical kinetic energy, this cornerstone of physics, a thousand times experimentally verified over the last four hundred years, is proclaimed as invalid by Einstein in STR.

His math physical construction forgery allegedly proved that kinetic energy is not conserved at different inertial frames and that just momentum (linear summation of Einstein's newly faked up kinetic energy $E = mvc = pc$ with linear change of velocity) is conserved. This biggest fallacy of STR is for a hundred years dogmatically repeated without a single analysis and detection of this fraud. Connection of a contraction of the one dimension of bodies $l = l_o\sqrt{1-v^2/c^2}$ and rise in energies of these bodies $E = mc^2/\sqrt{1-v^2/c^2}$ in STR with the same Lorentz factor is fatal mistaken and means returning the physics to Descartes physics based on mv which was denied at the end of the 17th century.

This mistake is later re-manipulated in 1907 by Minkowski in his space-time $x^2 = c^2t^2$ followed in 1908 by David Hilbert at construction of Hilbert space $\mathord{\mathrm{I}}^2$ as $\int dx \left| \mathord{\mathrm{I}} \right|^2$ which was used by Einstein in GTR (thus he returned to classical kinetic energy) despite Einstein's prior faked up mathematical proof in STR that kinetic energy is not conserved in inertial frames.

120

Einstein's mathematical forgery in STR at misusing Taylor's series for Einstein's return from $E = mc^2 / \sqrt{1-v^2/c^2}$ to reality of classical kinetic energy $1/\sqrt{1-x^2} = 1 + x^2/2 \ldots\ldots + 3x^4/8 + 5x^6/16 + \cdots$ is dogmatically repeated for hundred years and was never parsed. For both mathematical relation $1/\sqrt{1-x^2}$ and $1/\left(1-x^2\right)$ using Taylor's series we can obtain the same result $1/\left(1-x^2\right) = 1 + x^2 \ldots\ldots + x^4 + x^6 + \cdots$ or also

$$1/\left(1-(x/2)^2\right) = 1 + (x/2)^2 \ldots$$

Energy for reliance on velocity rises as $E = \gamma^2 m_o c^2 = m_o c^2 / \left(1-v^2/c^2\right) = m^2 c^2$ and relation $E = mc^2$ is a blunder or just a first approximation.

If Einstein had really wanted to establish an equation between own energy and mass of material objects he would make it equally with electrostatic, where own energy of electrostatic source is $W \approx Q^2$ which results from the conservation of Coulomb forces around the source. Also own energy of the source in today's static gravity field theory is $E \approx m^2$. The result is that classical kinetic energy in fact represents $E = m^2 v^2 / 2 = m^2 c^2 v^2 / 2c^2 = p^2/2$. At low velocities compared to c change of mass m is neglected in classical momentum $p = \gamma m v \approx m v$ as well as change of energy corresponding to this change of mass is neglected in classical kinetic energy $E = \gamma^2 m^2 v^2 / 2 \approx m v^2 / 2$.

In conclusion, we may say that regardless of whether the Michelson–Morley experiment provides zero or other results, regardless of whether the ether or gravity waves exist or not, regardless of whether the Universe is static or is expanding, whether light is bending in gravity, whether black holes exist or not Einstein's new physical laws or the modification of previous existing laws in physics are wrong and incompetent. Existence or non-existence of ether, black holes, gravity waves, expansion of the Universe do not confirm the correctness of Einstein's mathematical illusion of new physical laws and their teleological fraudulent approximation.

5. The Intrinsic Mission of Time in Physics

The physicist that seeks to seriously ponder what represents a quantity of time in physics, may spend any time figuring this, but in the end the man must come up with only a single answer. This answer is the same as Aristotle's that time is the measure of the speed of movement. The same answer attributes to physicist Julian Barbour who, after a 50 year inquiry into what time is in physics, came to the conclusion that *"Time is nothing but a measure of change and time itself does not exist"*. It is pity that this correct finding brought him to metaphysic of nihilism that the motion is pure illusion. *"I believe that time does not exist at all, and that motion itself is pure illusion."*

If there is no movement of objects, no time or velocity exists. It is the only existing change we are unable to assign any time or any velocity or any acceleration. For two mutually moving objects we are unable to assign any speed and we need a third comparative calibration of speed in order to do so. Time, velocity and acceleration represent the comparison of the count of speed of one change to another.

Time in contemporary physics is not an arbitrarily chosen variable, which could by itself span, lapse and vary independently of objects. The basic concept of modern physics for the last four hundred years lies in the fact that, by establishing basic units of length and time, at the same time, the basic unit of uniform velocity is defined as the ratio of this unit length and unit time. The calibration values of all fundamental physical constants are based and firmly linked to this basic definition. By this definition, the concept of unit time in physics is established as the speed of movement on a defined distance in the space, whereas at once the basic calibration comparative value of velocity is defined.

If we want to measure any quantities in physics we need the calibrated gauge to do so. For distances we have 1m ISO gauge, for mass 1kg gauge, for temperature we have calibrated thermometers and so on. But where is the gauge for very primordial quantity in physics - movement or velocity as its scale? Time is not a non-material quantity which is measured by ticking of a clock but ticking of the clock, 1second, fixed to 1 meter is non-conscious calibrated gauge of velocity for comparing two movements. The change of the time unit is the change of the basic comparative unit of velocity- change of the basic gauge of velocity.

So the unit time and unit velocity are firmly fixed and represent the same in inverse proportionality - basic comparative speed of movement. If a body moves ten times faster than the unit velocity, it then travels ten times more unit length in one unit time or travels one unit length in one-tenth of the unit time.

We can express the velocity identically as the ratio of travelled unit length to the unit time, or the inverse ratio of elapsed unit times to length unit. We do so also in many practical situations. The speed of the runners, e.g. at a distance of hundred meters we express by the ratio of ran-off times. The acceleration of cars we express by the ratio of elapsed times at the fixed distance.

Einstein never understood that for each measuring of velocities alongside the primary reference frame of the observer another two reference frames are required. The second moving reference frame is that with calibrated unit velocity movement and the third reference frame is the measured frame whose velocity of movement we want to measure. In rigorous physical experiments we would have to compare calibrated movement of unit velocity with unknown velocity which we want to measure. Instead of calibrated mechanism with moving unit velocity we use a clock (one second) which serve us as non-conscious calibrated gauge of unit velocity for comparing velocity of another movement.

The fraud of Einstein's time dilation at explaining the M-M experiment can be manifested on the following example, understanding to each man with elementary education.

If a car is moving at a distance 10m with velocity 1m per second than it takes 10 seconds this car to travel the distance 10m. This time 10 seconds is the ratio of distance 10m and velocity 1m per second. Time is the result, the consequence of this ratio. The unit time and unit velocity are firmly fixed with basic definitions of physics and the pertinent size of the change of the unit time is firmly directly fixed with the same size of the change of the unit velocity.

If in this case not the ratio 10 seconds is measured than just two or one of two explanations is possible. Or the distance is not 10m or the velocity is not 1m per second or both are not correct. However, Einstein claimed (that's how he explained the M-M experiment) that this distance and velocity are correct, but time, unit time 1s, has changed (dilated), not as the precision of time measurement nay as the change of distance or velocity, not as a quantity fixed firmly to unit velocity, but as a new fancy quantity fixed with velocity at a new relation $t = t_o / \sqrt{1 - v^2/c^2}$. Einstein by this 'new conception of time' from his Dreamland of physics replaced the physical reality of the wrong given velocities, their unknown direction and from this resulting also the improper distances in the interferometer at the explaining of Michelson-Morley experiment in the beginning of the 19th century. So the causes were exchanged by consequences in metaphysics of time.

Today we know that the velocity of the Earth (it moves not only around the Sun but at least also around the center of Milky-way galaxy) in the surrounding space, their mutual direction (so also the distances) so basic suppositions at explaining the M-M experiment until 1919 were wrong.

In 1983 (17th CGPM) a length of 1m was defined as length of the path travelled by light in a vacuum in $t_c = 1/299\ 792\ 458$ second. By this definition the ratio of unit velocity, unit length and unit time compared to velocity of light in vacuum was inseparably fixed. The unit of time, 1 second, is the speed of the movement measured on unit length, which in comparison with the speed of light is 299,792.458 times slower.

The basic relation of STR v/c for ratio of velocities can be identically expressed as an inverse relation of times $v/c = t_c/t_v$ where ratio of two velocities or inverse time is compared by third- basic unit velocity or unit time. The unit of time of one second then does not represent the ticking of the clock but the basic comparative speed of movement of the body for comparing other speeds on the length of one meter.

In STR for the comparative speed the speed of light is selected, which is constant in all inertial frames. Light moves on one meter in the system of the observer or on the contracted meter in the moving frames at the same speed and thus a unit of time per 1 second is also established and fixed in all STR inertial frames.

If, by changing the speed of the inertial system, the unit of length contacts in STR as $l = l_o\sqrt{1-v^2/c^2}$ then to ensure the validity of international SI definition of 1 meter from 1983 and ensure a constant velocity of light in all inertial frames as well as ensure validity of the calibration value of all basic physical constants we must also contract unit time as $t = t_o\sqrt{1-v^2/c^2}$.

Time dilation $t = t_o/\sqrt{1-v^2/c^2}$ fixed firmly with length contraction $l = l_o\sqrt{1-v^2/c^2}$ in accordance with STR leads to the disintegration of the calibration value of all basic physical constants.

So if unit time in STR changes as $\sqrt{1-v^2/c^2} = t_v / t_o$ than basic comparative unit of velocity change as $1/\sqrt{1-v^2/c^2} = t_v / t_o = v_o / v_v$. So if STR claims that in STR time dilates and ratio $v_o / v_v = 1$ is kept the same then also the basic comparative unit of speed as well as speed of light in inertial frames with increasing speed is slowing down (for speed of light is also valid 1c+1c=1c) so also dilates.

So for keeping constant speed of light in any inertial frames any ratio of changed lengths and changed times must be always constant in any inertial frames. If length 1m contracts to half meter so in order the speed of light remains constant the ratio of unit length and unit time must remain the same and 1 second has also be contracted to half second. If STR claims that when length 1m contracts to half meter then 1 second dilates to 2 seconds then the speed of light falls to one quarter. In fact, but in STR basic comparative speed unit has changed four times.

Einstein's concept of time as an arbitrarily chosen variable determined at each point of space, ticking of clocks, which could by itself span, lapse and vary independently, without fixation of the time and the time unit to the specific moving object has to be labeled as naive and Dadaistic. This metaphysics conception of Einstein's can be clearly seen e.g. in his 1907 paper [12, V2, D47] or in his 1910 paper [12, V3, D2]:

"To determine the time at each point in space, we can imagine it populated with a very great number of clocks of identical construction in rest frame. The totality of the readings of all of these clocks in phase with one another in phase is what we will call the physical time. By the time coordinate of an elementary event we will understand the indication of clock that is situated infinitely close to the point at which the event takes place."

For Einstein, each uniformly moving frame has its own time depending on the velocity of the frame.

On this, Einstein's fantasy land (as called by Schwarzschild) or dreaming up fields (as called by Lenard) is erected the metaphysics of time dilatation or space-time mathematical constructions of Einstein's *Special and General Relativity* which has no connection to physical reality. Transition to reality from Einstein's metaphysics is attained in STR and GTR by misleading, fabricated, teleological approximations and in this manner the metaphysics of STR and GTR is allegedly confirmed.

Explaining the physical phenomena using time dilatation as e.g. in M-M experiment or using space-time continuum as a gravity field is devastating for exploring and understanding the physical picture of the Universe.

The most flagrant example of the fall of physics at the nescience and darkness is the metaphysical explanation of the alleged longer lifetime of fast moving muons compared to stationary muons in using time dilation or vice versa to submit this physical phenomenon as the evidence for the confirmation of STR validity.

A great number of experiments observed the decay of muons after passing through the atmosphere of Earth (or other various materials as iron, lead or plastics) were carried out. It is necessary to stress that in all these experiments the decay of the muon is a consequence of energy loss when passing through the material. The so called dilatation of the time represents the different periods for high velocity muons in losing their energy in material passing through.

Ensuing one of the real physical explanations of muon lifetimes can be drawn from the results of this book or previous works of the author.

There is no doubt that in the vicinity of any elementary particles of matter (and therefore around each ponderable body composed of this elementary particles as called gravity), there is its own force field. This force field in the particle's vicinity is a result of particle creation, which is the spin product of a curl compression of ether infilling all space of the universe. The particle as a curl compression of ether and the force field in its vicinity as the gradient of ether are thus inextricably bound. This field acts on the particle surface by the opposite force pressure to the particle's internal centrifugal force of angular momentum, arising from the spin of the particle.

When the velocity of the muon is high, the pressure of its own field on its surface is sufficient for its stability. When the muon passes through the atmosphere, its velocity is reduced, its energy is passed to the environment and the pressure of its own force field on its surface is reduced. When the pressure is not sufficient for the muon's stability, energy in the form of neutrinos are passed to the environment and the muon is observed with minor mass as the electron.

Instead of this or eventually similar (change of density of passing material in different high of gravity field,...) actual physical causes of muon lifetimes, current physics provides the only explanation in the form of metaphysics and mystery of time dilation.

en.wikipedia.org/wiki/Time_dilation_of_moving_particles –
"*Moving muons should have a longer lifetime than resting ones as predicted by special relativity. The proper time of a clock co-moving with the muon, corresponds with the mean decay time of the muon in its proper frame. Therefore its proper time is shorter*".

As a similar example to the muon case, we can imagine a case of the meteor approaching the earth with small velocity that burns at higher altitude above the ground than the meteor approaching the earth with greater velocity that pass longer path and burns at lower altitude above the ground. An explanation of the reason of this longer path of the meteor in the atmosphere of the earth by the time dilation of a clock connected with this meteor or more over take this as the prediction or the confirmation of time dilation in relativity is similar rubbish as this construction in the case of the muon. (As is detailed below we see here Maupertuis metaphysics of today's physics in full action)

Outside of the time dilation, current physics provides no other explanation of muons 'lifetime' change with reliance on length path in the atmosphere. Current physics is completely satisfied with this explanation and any other explanation is unwanted.

There is no doubt (confirmed also by Einstein after 1920 when he rejected his Special theory of relativity) that velocity is not a causation of the length contraction of bodies or alleged time dilatation in STR. Purported time dilation in muon decay as well as in M-M experiments or curvature of space time at gravity fields is a consequence of factual physical causation of material substances. In special and general relativity, the physical causation (of factual change in densities and pressure of material world) is replaced by its consequences in metaphysics of time, where the consequence is proclaimed the cause of the physical processes.

In physics, concepts of time and movement are identical concepts, as are also the unit time and unit velocity identical concepts.

Einstein does not differentiate between the unit time as the calibration gauge of movement, so unit velocity as the scale of movements, and the time as the time duration of movement measured by this gauge or scale. These two notions are packed in one, mingled in his mess of simultaneity and are the source of never ending contradictions in STR. Einstein's absurd invariant requirement on the invariable traveling time on different length of path or the same traveling time with different velocities on the invariable path is, in STR, fulfilled with time dilatation. This is but the same as change of calibration gauge of movement or change of unit velocity as the scale of movement. By these procedures the absurdity of constancy of different velocity to velocity of light are manipulated.

As the basic calibration unit velocity we can in physics arbitrary correctly choose the velocity of 1m/s or the velocity of light. If we chose the velocity of light as comparative velocity and if we simultaneously declare that all velocities to velocity of light are always the velocity of light (as it is declared in the second of two postulate of Einstein's special theory of relativity), then no other velocity than the velocity of light we are able to measure. As a result of this antagonism, de facto all physics is destructed. However, such a theory is declared by mainstream physics and academics all over the world as allegedly the greatest intellectual achievement in the history of mankind.

For the reason of lack of understanding of the concept of the time in physics, Einstein's special and general relativity must be rejected from physics.

6. Distortions and Defects of Quantum Mechanics and Quantum Physics

The principle of least-action is the central principle of QM and was also fully applied by Einstein in his relativity [e.g in 12, V2, D47, p. 300, 1907]. In the principle of least-action, the variational principle introduced by Maupertuis in 1747 is used to find the shortest path or 'least time' to obtain the equations of motion for the system. In the principle of least-action, the physical cause or any material physical phenomena responsible for movement of bodies are suspended. This principle becomes more and more a central principle of today's physics to derive the QM and Relativity equations (and even the equations of classical physics).

In 1746 Maupertuis wrote the work - Derivation of the laws of motion and equilibrium from a metaphysical principle - with two head chapters – I. Assessment of the Proofs of God's Existence that are Based on the Marvels of Nature, II. Need to Identify Proofs of God's Existence in the General Laws of Nature

Let us recall the E. Mach judgment on this principle [31] in 1919.

"Maupertuis, in 1747, announced a principle that he called the principle of least-action. He declared this principle to be the one that eminently accorded with the wisdom of the Creator.

He took as the measure of the 'action' the product of the mass, velocity, and space described, or mvs. Why, it must be confessed, is not clear. By mass and velocity definite quantities may be understood; not so, however, by space, when the time is not stated in which the space is described. If, however, a unit of time be meant, the distinction of space and velocity in the examples treated by Maupertuis are, to say the least, peculiar.

It appears that Maupertuis reached this obscure expression by an unclear mingling of his ideas of vis viva and the principle of virtual velocities. ....It will thus be seen that Maupertuis really had no

principle, properly speaking, but only a vague formula that was forced to do duty as the expression of different familiar phenomena and not really brought under one conception It would seem almost as if something of the pious faith of the church had crept into mechanics".

Since this book is written not only for physicists, active in the topic of QM but also for the wider physical community or non-physicists, the equations in this section are written not in their full rigorous mathematical form but rather as equations in their most simplified form manifesting their physical concepts to the general public. For their rigorous mathematical form see [10]. But due to the results of this work, that classical kinetic energy in fact represents $E = \gamma^2 mv^2 / 2 = m^2 v^2 / 2 = p^2 / 2$ and so the central auxiliary formula of the 20th century $E = p^2 / 2m$, used as the default relation in QM, is not valid. This simplified form is thus many times more correct then those rigorous ones.

If we seek to evaluate a chapter of physics called quantum mechanics (QM), we must indicate what QM is.

Terms 'Quantum mechanics' or 'Quanta' originate from the notion of quantization. Quantization simply concerns the fact that the rigid matter of nature is formed from separate units - atoms and molecules. In the vicinity of nuclei (clusters of protons and neutrons) of these units there co-exist radial gradients of positive electric fields, shielded by layers of electron shells. Confirmed quantization or quanta in physics simply relates to the non-continuous spectra of electromagnetic radiation; photons, emitted by electron shells with different energies from different levels in gradients of these electric fields.

Today we can treat electrons in an atom as the field of spinning sphere shells, with corresponding thickness where quantization means that no two shells can naturally concur or 'occupy' the same space of shared sphere shells.

But if electrons are free, outside of atoms and if there is no space restriction (restriction on dimension of electrons shell thick) of radial fields of atoms, electrons (or protons or any charged particles) show no restraint, no quantization in their energy spectra as was experimentally demonstrated (in period from 1909 till 1960) by their continuum energy emission over the entire electromagnetic spectrum at braking radiation., synchrotron radiation, cyclotron radiation or at beta decay.

Similarly, except for the fact that photons are individual unit entities, no restraint and no quantization of photon energies was ever observed. Energies, frequencies and wavelengths of photons in relation to the proportionality of energy $E \approx h\nu \approx h / \lambda$ are produced continuously from zero to infinite values.

No smallest limit of the energy of electrons or photons was ever observed as is misleadingly declared in quantum mechanics in the vision of the obscure enigmatic Planck constant. Certainly no such smallest limit (in fact a smallest limit of ether medium) until today has been discovered [10, 11].This is the content of Feynman's declaration: *"We do not have a picture that energy comes in little blobs of a definite amount".*

These experimental facts disproved the validity of the initial physical declaration of Einstein on the first page of his article on the photoelectric effect from 1905 awarded by Nobel prize - *On a **Heuristic Viewpoint** Concerning the Production and Transformation of Light* [12, V2, D14] - *"The energy of a ponderable body cannot be broken up into arbitrarily many arbitrarily small parts, while according to Maxwell's theory the energy of a light ray emitted from a point source of light spreads continuously over a steadily increasing volume."*

In 1907 Planck warned the young patent clerk that he had gone too far, and that quanta described a process that occurred during emission or absorption, rather than some real property of radiation in a vacuum. *"I do not seek the meaning of the 'quantum of action' (light quantum) in the vacuum but at the site of absorption and emission"* [44].

But this Einstein's work ("heuristic viewpoint" means idea without evidence) was awarded a Nobel prize. Striking is the fact that Einstein did not deliver a conventional Nobel lecture, the summary results of his work on photoelectric effect, before or on the occasion of the Nobel Prize Ceremony. Although being noticed staying in Stockholm, Einstein chose to travel to Japan and did not attend his Nobel Prize Ceremony in 1922.

Quantities joined by proportionality of Planck constant, e.g. energy and frequency, runs continuously from zero to infinity and thus also the value of these quantities are continuous and limitless.

Planck himself failed to find any physical significance for his constant h, which he called 'quantum of action', beyond its appearance in the radiation formula, despite spending many years trying to do so [57]. Planck's message states, *"The thermodynamics of radiation will therefore not be brought to an entirely satisfactory state until the full universal significance of the constant h is understood"* [57].

In our previous paper [10], [11] it was shown that the ratio of the Planck constant and light velocity represents the basic density of momentum and pressure of a flat ether medium in void space.

All the so called Base Planck units as the combination of Universal gravitational constant G, the speed of light in vacuum c and Planck constant h - Planck length, mass, time, charge, temperature- are Planck's baseless fantasies of Einstein's kind. Universal gravitational constant G is not a constant but it is a calibration value that corresponds to an arbitrarily chosen by man amount of masses (1kg). This 'constant' is changing if we chose different amounts of masses for its measurement. If we suppose the constancy of light velocity in vacuum c and constancy of Planck constant h then all so called Base Planck constants change with this change of G.

For the same reason, Einstein's gravitational constant as a combination of G and c is also a baseless fantasy and Base Planck units, as well as Einstein's gravitational constant, must be rejected from physics.

Quantum mechanics is a procedure that attempts to describe the motion or motion-states of fundamental particles of matter (primary electrons) and transfer of energy at changes of motion-states into environment in mainly two situations - a central force field in the vicinity of different atoms and in free movement without the action of external forces.

In the case of the motion of the electron in the central field of the protons for the hydrogen atom (two-body problem), physics (now called classical physics) satisfactorily provided (in the presented relationship of Bohr and in the first presented relationship of Schrödinger) an explanation for the amount of energy needed to be added or removed from electrons (the spectral lines of hydrogen) for the occurrence of electron at different distances from the center of the proton.

Today we can treat electron in atoms as the spin of sphere shell field with corresponding thickness where quantization means that two shells cannot naturally concur or 'occupy' the same space of a shared shell.

Subsequently, in other cases than the hydrogen atom, it would necessarily have been stated that the force fields around the nuclei of atoms composed of a large number of nucleons have a complex character (and so also their spectra as the energy states of electrons). Also, in classical physics, we can't even satisfactorily solve the three-body problem analytically. Moreover, as was learned later (after 1930), nucleons in nuclei at least spin, if not spin at shells and, in addition, the composition of the atomic nuclei from protons and neutrons was not known by 1932.

From a global perspective, quantum mechanics can be characterized by two fundamental distortions. The first is the deformation resulting from the impermissibility of refusal or even opposition to Einstein's linear relationship for energy, with frequency of the photon after which energy matches photon momentum. The impermissibility of this refusal is overcome in QM by introducing a mysterious de Broglie's wavelength of matter and Schrödinger's wave function.

Secondly, it is a deformation resulting from the impermissibility of refusal of Einstein's claim about the absence of ether and the inevitability of its repeated introduction in quantum mechanics in the concept of the energy of vacuum by providing it the physical properties equivalent to ether in electrodynamics.

Schrödinger came out in his first paper on undulatory theory of mechanics in 1926 [20] from a basic physical supposition - *"The wavefunction physically means and determines a continuous distribution of electricity in space, the fluctuation of which determines the radiation by the laws of ordinary electrodynamics. In the case of the hydrogen atom, it has been possible to compute fairly correct values for the intensities e.g. of the Stark effect components by the following hypothesis: the charge of the electron is not concentrated in a point, but is spread out through the entire space proportional to the quantity $\psi\psi$. The fluctuation of the charge will be governed by ($\psi\psi$) applied to the special case of the hydrogen. To find the radiation, that by ordinary electrodynamics will originate from these fluctuating charges, we have simply to calculate the rectangular components of the total electrical moment integrating ($\psi\psi$) over the space."*

For the Atomic Spectra of elements other than the hydrogen atom with the larger number of nucleons in the nucleus of an atom, Schrödinger's procedure failed to provide a satisfactory value. Instead of considering that the electromagnetic field around atomic nuclei other than hydrogen atoms are complex and hitherto unknown (eke unknown spin 1930 and neutron 1932), during the formation of quantum mechanics 1924 -1930 continued the hunt for provisions of mathematical constructions describing the spectra of atoms.

In fact, the spectra of atoms show us how the electromagnetic fields, gradient of ether, around nuclei of these atoms look and so the most physically natural approach would have been in an effort to model this field by the laws of ordinary electrodynamics.

But ether was slain and banned by special relativity and gotten rid of by general relativity, where force fields turned into the curvature of non-material quantities of space and time. Schrödinger's good-will for mathematical structure associated with specific physical realities was dismissed (shortly also by Schrödinger). Theorists Born, Heisenberg, Jordan, Hilbert, Wiener, Pauli, Eckart, Kramers, Dirac, Sommerfield, Weyl, Neumann and Wigner bred fictive, bizarre mathematical structures which combined the physical unknown go-as-you-please quantities, variables and parameters.

Conceptions such as operators, matrices, extra matrices and continuous matrices, commutators and anti-commutators, approximation, group methods and symmetries, frequencies, wave lengths, wave functions, relativistic and non- relativistic corrections, delta functions and coupling constants were incorporated in various - statistical, probabilistic, uncertainty, energetic, time, momentum - interpretations of quantum mechanics. These go-as-you-please quantities, variables and parameters have mostly no connection to physical reality and so, to this day, no one understands quantum mechanics and nobody knows how the particle moves in a field described by quantum mechanics.

Noteworthy is the Dirac attempt to link his theory with physical reality, which arrived with the statement (known as the Dirac sea) that the whole universe is filled with anti-electrons. This statement was factually the same as Schrödinger's prime supposition.

In 1927 Ehrenfest (who was, according Einstein himself, merely the best teacher in our profession whom he have ever known) in his theorem linked the classical and quantum pictures without approximations at declaration that the expectation of quantum mechanics is equal to the expectation value of the negative gradient of the potential function equivalent to Newton's second law of motion.

The beginning of the formation of wave and quantum mechanics mainly connects with the names of de Broglie and Schrödinger. In the case of de Broglie, although Einstein was not supervisor of the de Broglie doctoral thesis (substantial work of de Broglie, 1924), Einstein led de Broglie through the steps of his work. De Broglie's final version of his doctoral work was even sent to Einstein for approval. Without this approval, the defense would not have been accepted. De Broglie, in a preface to the German translation of his doctoral thesis stated: "*Einstein from the beginning has supported my thesis*".

Afterwards, Einstein praised de Broglie's work in his letters to Lorentz, Langevin and other physicists [12, V14, D385, D398, D399] - "*De Broglie's dissertation is a very significant writing. De Broglie raised a corner of the big veil. I have also found some things supporting his construction*".

Einstein with the help of de Broglie installed into quantum mechanics Einstein's construction [introduced in 1907 paper 'On the relativity principle and the conclusions drawn from it' ,12, V2, D47, p. 268] of group velocity and dependence of group velocity on a frequency of the light or any rigid body transferring information by vibration (Einstein's metal strip).

The form of the Einstein's relation for group velocity is the same as the form of the already absurd relation for velocity addition. This absurdity is afterwards further extended so that for mutual velocity of light (or any bodies moving with light velocity) and body is valid the STR velocity addition relation but this velocity taken from frequency (number of wave in time) of this light is again not valid. Resultant velocity from addition velocity relation is again changed as the ratio of resultant velocity from addition relation and frequency and this new velocity is called group velocity. So for Einstein there exist two different mutual velocities of light and body.

Exactly this Einstein substantiality is the content of de Broglie's dissertation work concerning of two mutual velocity (phase and group) between two rigid bodies or between the faked up wave length of one body (in infinity in rest state) and the light. The de Broglie wave length of each rigid body which is spread in infinity in rest state and which yet at velocity 1m/s of bodies is contracted with limitless velocity from infinity length to a length less than 1 millimeter is allegedly physical reality for each rigid body, which is confirmed by the Nobel prize award for de Broglie in 1929.

In the case of Schrödinger, his written thanks to Einstein, which states that the formation of his equation would not be possible without the decisive contribution provided him by Einstein speaks for itself. Schrödinger heard of the de Broglie work while reading one of Einstein's papers and was intrigued by the concept of these so-called de Broglie waves. Thirteen letters between Einstein and Schrödinger were exchanged in a period of one year before the Schrödinger allegedly guessed his 'ingenious' equation on his holiday trip in December 1925. This 'ingenious' (rather rubbish) equation was published in his first paper on Quantum mechanics in 1926 [20] and afterwards during the next 25 years Schrödinger published a great number of papers on an explication of what physical reality represents in his undulatory theory of mechanics in Quantum mechanics.

On the initiative of Einstein in 1927, Schrödinger moved from Zurich to Berlin and there he became a colleague of Einstein's. He had been elevated to a professorship at the University of Berlin in 1927, just one year after introducing his first wave equation for quantum mechanics.

That this Einstein contribution, labeled by Schrödinger as decisive had rather a form of instructions can best be seen from Einstein's letter to Schrödinger on 31 May, 1928. It is not surprising (as is shown in this book) that the content of the instruction in this letter was the relationship between energy and the frequency – "*I say: not E and ν but rather: E or ν; But I cannot make head or tail of it mathematically*" [45].

As the relation between energy and frequency is fully fixed with the relation between mass and energy $E = mc^2 = h\nu$, this Einstein reasoning shows that both relations are simply Einstein's wish. Einstein had no physical reasoning for justifying this relation and, as we show in this book and previous papers [10, 11], both relations are wrong. These two equations are, for the last 100 years, the basic law of physics. For the last 100 years the experimental results of thousands experiments are expressed in so called energy units using these equations, although even such physicists as Feynman labeled the energy units so also equations $E = mc^2 = h\nu$ as idiocy – ***"Idiocy of energy units – It's too bad, but I have already apologized and there is nothing more I can do."*** In detail, see authors paper [11, 10].

Einstein's equations for photon energy $E = h\nu$ and photon momentum $p = h / \lambda$ from which arise relation $E = h\nu = h\nu / c \cdot c = h / \lambda \cdot c = pc$ are primary physical relations in physics until today. These relations are the primary physical relationships that caused the degeneration of the physics of the 20 century [10], [11]. In 1900, Planck, in accordance with then still valid scientific principles, carefully declared that photon energy can be considered proportional to the frequency of a photon $E \approx h\nu$. Einstein, without any experimental evidence, in 1905 made a 'big scientific discovery' when he simply declared $E = h\nu$ and based STR also on $E_k = pc = mvc$.

For confirmation of this 'discovery', Millikan carried out the experiment in 1914. With all respect to the greatness of Millikan's physical experimental skills, he succumbed to the pressure of the power structures and agreed with their interpretations concerning the validity of the linear relationship of energy on the frequency of a photon in his experiment. This agreement by Millikan was a condition of the Nobel Committee for the award of the Nobel Prize to Einstein in 1921 for the photoelectric effect. In 1921 Millikan became director of the laboratory at CalTech and won the Nobel Prize in 1923.

Millikan in his experiment allegedly confirmed for electrons proportionality $E = h\nu = eV = \frac{1}{2}mv^2 = p^2/2m_o$ from which arise $p \approx \sqrt{h\nu} \approx \sqrt{h/\lambda}$ what contradicts $p = h/\lambda$ so that fully disproves Einstein's $E = h\nu = h\nu/c \cdot c = h/\lambda \cdot c = pc \neq \approx \sqrt{h\nu} \cdot c = \sqrt{hc/\lambda} \cdot c$. Also Schrödinger's $E = h\nu = p^2/2m_o = h^2/\lambda^2 2m_o$ was in full conflict with Einstein's $E = h\nu = h/\lambda \cdot c = p \cdot c$ therefore the introduction of the cryptic wave function ψ by Schrödinger was necessary.

In this way Schrödinger sought to overcome the absurd logic of the Einstein's relationship between energy and the momentum which can be best documented with Einstein's work from 1916 [12, V6, D38]. In this work Einstein claims that momentum of a photon is $p = h\nu/c$, that this momentum multiplied by c is its energy $E = h\nu$ and he also claims that momentum squared is $p^2 = h^2\nu^2/c^2$. Here we clearly see that for Einstein in his fantasy world $E^2 \approx p^2$, although in the world of real experimental physics $E \approx p^2$.

Opposition to these relationships was (and still is) not permissible. So de Broglie and Schrödinger introduced mystical physical non evincible quantity (dimensionless point particles connected with wavelength in infinity or wave function) employed through obscure operators in constructed robust mathematical theories, so that using them could accommodate experimental data and at the same time keep Einstein's relations valid. From the physical point of view these theories are, even for top specialists, beyond all understanding.

R. Feynman (the Nobel Prize laureate for quantum mechanics! 1966): *"I think I can safely say that nobody understands quantum mechanics"*. *"We have always had a great deal of difficulty understanding the world view that quantum mechanics represents (1965)."*

S. Weinberg (the Nobel Prize laureate in particles physics! 1979) in 1992: *"I am a little uncomfortable knowing I had spent life working a field that nobody understands. How could all the questions about quantum physics have been answered if it is still not understood by its practitioners?"*

Einstein's degeneration of the relation of physical quantities of momentum and energy for photons and, as was shown in [10] also for mass bodies, has become the default physical premise of the theory of special relativity and quantum mechanics, linking these physical quantities $E = pc = (h/\lambda)c = mvc$ which then differs only by the constant c. For relation of photon energy $E = h\nu$, (that in the form of differences of energies $\Delta E = h\nu_2 - h\nu_1$ at photoelectric effect was awarded by the Nobel Prize), however, neither Einstein nor physics up to today has told us what physical reality represents the Planck constant itself (action of what it is) and what is the frequency of a photon or what the physical properties of the photon we have to the frequency of the photon assignee [10], [11].

The frequency of a photon cannot in any way be measured. We can measure the wavelength of the photon and, in relation $\nu = c/\lambda$ tie the photon wavelength to its unknown physical quantity frequency. Then for Einstein's explanation of photoelectric effect the difference in momentums $\Delta p = hc/\lambda_2 - hc/\lambda_1$ in fact primitively explains this effect.

De Broglie (1924) founded his theory on the Maupertuis least action principle $mv\delta s$ and extended this deformation with the dispersion relationship of frequency and wavelength for material particles at the group velocity relation $w = \Delta\nu\Delta\lambda$. Hereat, according to de Broglie, all at once the relation (later Schrödinger's equation) $\Delta\nu = n/\Delta\lambda^2$ must be valid. But these fictitious mathematical constructions of frequency and wavelength can't anyhow be experimentally measured. Nobody up to today knows what physical properties represent the frequency of particles and don't even know what is the wavelength (defined in infinity) of these particles.

In physics textbooks it is stated that Davisson and Germer, in 1927 in their electron diffraction experiments, confirmed (it was their wish to peer to the Nobel Prize) the de Broglie hypothesis so also that electrons in rest state have wave in infinity. At first glance, it's obvious that this allegation is unfounded and tendentious. If the so called waves of matter were confirmed and, on top of that even in 1929, awarded by a Nobel prize there would be no troubles for de Broglie himself and thousands of succeeding physicists continuing up to today with what reality of wavelength of mater is and what physical reality QM represents- see WIKI -Louis de Broglie *"In his later career, de Broglie worked to develop a causal explanation of wave mechanics, in opposition to the wholly probabilistic models which dominate quantum mechanical theory"*.

The explanation of observed constructive interference picture at the Davisson–Germer (1927) diffraction experiment with electrons lies rather in superposition (in more detail referenced below at Double-slit experiment) of recoiled paths from and curved paths around atoms of nickel crystal than in the confirmation of de Broglie's mysterious waves defined in infinity for the rest state of bodies.

But first of all the extension of de Broglie's idea from an electron whose form structure is unknown (it most likely has no definite form and in atoms forms a force field at shells, free electrons in conductors forms planar surface force fields and while emitting from atoms or outside of atoms and conductors occurrence, we may think, it forms a spherical object) to all other rigid bodies is false and pure baseless fantasy.

The principle of de Broglie's construction of wave of matter for particles or macroscopic body moving at velocity v in relation $mv = h / \lambda$ is physically wrong, because the movement of one body we can consider as the movement of more parts of this body together or as several smaller bodies bound to each other. Waves of the macroscopic body as a whole have microscopic value, but each of its divided smaller parts (e.g. atoms of the same body) has wave growing to infinity.

In our previous paper [10], [11] it was shown that the Planck constant must bind with basic flat density and pressure of ether in void space. Subsequently, in relation $h / \lambda = mc$ wavelength λ constitutes the dimension of curl compression of the photon momentum, as well as a dimension of localization of spinning internal momentum of rigid elementary particles at rest.

Thus the correct de Broglie consideration about moving rigid particles at velocity v (in comparison to h^2 / λ^2 basic calibration 'rest' state) represents [10], [11] the relation $h / \lambda - h / \lambda_0 = mc - m_0 c = mcv / c = mv = p$ constituting the responding contraction of each of the primary construction particles of macroscopic bodies. This contraction is a result of a change of pressure of the force fields surrounding particles (gradient of ether freely pervading interstices in matter and inseparable joined with particles of matter which are spin products of ether) on the surface of those particles, due to the change of velocity. Its consequence is a contraction of the macroscopic body as a whole.

In the case of the classic experiment of QM, electron microscope, instead of assign to the dimensionless point electron obscure wave λ in fact λ constitutes the dimension of localization of electron. The greater is the velocity of electron so much greater is its energy which equals to the shorter dimension of electron localization λ .

Although Compton (Nobel prize in 1927), up until 1923, during 20 years of his experiments on the collision of photons with matter, used for the momentum of a photon the relation $p = h\nu / c = h / \lambda$ and for energy of a photon the relation $E = h^2 \nu^2 / c^2 = h^2 / \lambda^2 = p^2$ (known to Einstein at least from 1916 [12, V6, D38]), Schrödinger in 1926 arrived with another of the 'greatest achievements in the history of mankind' in his equation $\Delta E = \Delta h\nu = hc / \Delta\lambda = \psi h^2 / \Delta\lambda^2$ (1933 Nobel Prize). As reasoning for the foundation of this Schrödinger equation, physics and teachers at schools posit that this equation was guessed by Schrödinger's physical insight.

But so far physics do not identify any other relation between the wavelength and the frequency of a photon than $\nu/c = 1/\lambda$. So if Schrödinger realized that for energy spectra of atoms energy of photon represents h^2/λ^2 (what is in fact Bohr's or rather Rydberger's or Balmer's relation $1/n^2$) then the only connection to the relation between energy and frequency is $h^2/\lambda^2 = h^2\nu^2/c^2$. It is clear that the Schrödinger's effort (as he wrote "*would not be possible without the decisive contribution provided him by Einstein*") was to find the trick (in introducing the cryptic wave function ψ) to retain Einstein's faked up relation for energy of photon $E = h\nu$ as valid. The invalidity of this relation also explicitly means the invalidity of the relation $E = mc^2$ as they are firmly fixed in the relation $E = mc^2 = h\nu$ and also means the invalidity of all conclusions of STR.

Up to today nobody has yet told us what is the wave function ψ (as well as what is the frequency of a photon or Planck constant) and to what physical properties does it belong.

To date there are no universally accepted derivations of Schrödinger's equation from appropriate physical axioms just like Einstein's energy equation $E = mc^2$.

In 1926 the so called relativistic Schrödinger equation in a Klein-Gordon (K-G) form equation (falling short of a Nobel prize award) was presented as $h^2\nu^2/(m_0{}^2c^2)c^2 - h^2/(m_0{}^2c^2)\lambda^2 = -\psi/\psi$, from which follows [10, 11] that energy equals $E = h^2\nu^2/c^2 = h^2/\lambda^2 = m^2c^2$. So the correct writing of Schrödinger equation is $E = h^2\Delta\nu^2/c^2 = h^2/\Delta\lambda^2$ and no wave function ψ is needed.

Subsequently, in 1928 Dirac (1933 Nobel prize) presented the equation $h\nu/(m_0c)c - h/(m_0c)\lambda = -\psi/\psi$, in which nobody up to today knows what physical properties represents the frequency ν of particles, what physical properties represents the wavelength λ (defined in infinity), what is the wave function ψ or to what physical properties it belongs, what represents the Planck constant itself or action of what it is. From Dirac's equation it follows [10] that momentums $p = h\nu / c = h / \lambda = mc$ equal, so he at last allegedly reached (of course in advance dictated) the first quantum mechanics theory that fully accounts for special relativity.

Cultish equations for energy of photons $E = h\nu = (h\nu/c) \times c = pc$ and energy of particles $E = mc^2 = (mc)c$ or $E_k = (mc)c - (m_0c)c = mvc = pc$ glorified not by physicists, but by mass media, in which Einstein simply purposefully multiplied momentums by c together with parallel cultish Schrödinger equation $\Delta E = \Delta h\nu = hc / \Delta\lambda = \psi h^2 / \Delta\lambda^2$ were not rejected from physics to this day.

In our previous paper [10, 11] it was shown that, in fact wavelength λ represents the diameter of real dimensions of photons and particles as spin products of spherical curl compression of ether. Frequency ν represents the time of the light's flight through this diameter λ of photons and particles or equally their spin frequency. A decrease of diameter of the wavelength of the photon, a decrease of its volume, gives the increase in frequency of spin rotation of the photon that represents a change in the time interval of the light's flight through the dimension of the photon, all this while sustaining of constant surface rotation at the speed of light. Thus consequently, since a photon is an entity in quantum mechanics as well in Maxwell's electrodynamics, we receive for photon the same physical base in both theories as $\partial^2\varepsilon / \partial t^2 = \frac{1}{\varepsilon_0\mu_0}\partial^2\varepsilon / \partial r^2 = c^2\partial^2\varepsilon / \partial r^2$ so

$$\varepsilon^2 / c^2 dt^2 = \varepsilon^2 d\nu^2 / c^2 = h^2 d\nu^2 / c^2 = \varepsilon^2 / dr^2 = h^2 / d\lambda^2 = dm^2 c^2$$

and from the Pointing vector in electrodynamics, from Compton's works as well from the right hand side of Schrödinger equation we know that this writing represents writing for energy. So energy is proportional to frequency $E \approx h\nu$ but in proportion $E \approx h^2\nu^2 \approx h^2 / \lambda^2 \approx p^2$.

If we look into any experimental paper or textbooks in the field of particle physics e.g. [21] we find that relations $p \cdot p = m^2 v^2 = E^2 / c^2 = \gamma^2 m_o^2 c^2$ thus $p = mv = E / c = mc^2 / c = \gamma m_o c$ are valid relations for the actual physical phenomena.

With respect to momentum of photon $p = h / \lambda = h\nu / c$ Einstein's relations for energy of particles $E = mc^2$ as well as for photon $E = h\nu$ do not represent energy, but momentum intentionally multiplied by c. Energy of particles for reliance on velocity equals $E = m^2 c^2 = \gamma^2 m_o c^2$. So also, energy in relativistic mechanics represents relations $m^2 c^2 = m_0^2 c^2 + m^2 c^2 v^2 / c^2$ or $m^2 c^2 - m_0^2 c^2 = m^2 c^2 v^2 / c^2$ as equivalent relations to Klein-Gordon (K-G) equation or to the right hand side of the Schrödinger equation. So we receive the unified physical base (grand unification or theory of everything TOE) for unification of classical mechanics, relativistic mechanics, quantum mechanics an-d classical electrodynamics (and gravity as the superposition of remnants of electromagnetic fields around atoms) in kinetic (added) energy

$$E_k[2m_0] = p^2 = \gamma^2 m^2 v^2 = h^2 / d\lambda^2 = h^2 d\nu^2 / c^2 = \varepsilon^2 / c^2 dt^2 = \varepsilon^2 / dr^2 .$$

This unification has to be done for kinetic energy because classical and quantum mechanics does not know the concept of rest energy for free particles. These energies represent the amount of cumulative forces embedded into particles caused by change the speed v of particles or of bodies or by change the volume dr of particles, or energy embedded into ether at creation of photon of diameter of $d\lambda$ or dr

$$p = mv = \gamma mcv / c = h / d\lambda = hd\nu / c = \varepsilon / cdt = \varepsilon / dr .$$

We often hear the argument that the justness of application of quantum mechanics lies in the great accuracy of its math calculations.

But in fact, quantum mechanics so as with the Ptolemy epicycles, is a procedure for finding of mathematical construction to the existing experimental data which this mathematics has to arrive at. We can always find the mathematical relationships that describe experimental data or, if not, we can construct a new mathematics to describe the experimental data. But finding such mathematical constructions (theories) do not confirm the veracity of the basic physical principles (models) upon which these construction are based.

The Double-Slit Experiment and Casimir Effect

The double-slit experiment is another in a row of experiments for which explanation of the physics of the 20th century simply selects the most miraculous explanation. In 1801 a simpler form of the double-slit experiment was originally performed by Thomas Young. The double-slit experiment has later become a classic thought experiment, allegedly for its clarity in expressing the central puzzles of quantum mechanics.

In this experiment, the experimental data is accommodated by physically obscure and mysterious wave functions based on the mathematics of quantum mechanics theory. From this theory follows the mystery that particles or bodies can split and occur at two separate places simultaneously or the mystery that two particles separated by arbitrarily large distances can mutually communicate and transfer information with infinite velocity (the so called entanglement).

These mysterious explanations have subsequently for a hundred years been accepted by the power structures and mass information media, forcing the public as well as the wider physics community to accept them as the only possible explanation, although more simple, wise and feasibly reasonable physical explanations exist.

The primary physical condition of the double-slit experiment is that the sizes of two slits are equal to, or closer in size to the light's or electron's wavelength. If the slit's width enlarges (compared to the light's or electron's wavelength), the constructive interference picture becomes more and more unnoticeable.

In basic quantum mechanics textbooks, as well as in hundreds of papers and documentary films showing the alleged difference between the behavior of the classical objects and the mysterious behavior of quantum particles of matter, it is presented using the thought experiment, in which tennis or golf balls on a plate with two parallel slits are fired. The resulting image of two lines, where these balls supposedly fall is presented and is compared with the many lines

151

interference image that arises when the beam of light or electrons passes through a plate with double-slit.

But our physical thinking concerning the double-slit experiment must begin from the experimental knowledge of optics (since Newton) that light curves when passing closely around the edge of any object that we have at hand.

And so it is with the electron beam. The physical answer to why the electron path is curved passing at close proximity to the edge and path of a golf ball is not simple. The gravitational forces (identical with magnetic forces) at close distances (comparable to electron wave length - dimension of electron) around the material edges has sufficient power to cause the curvature of the electron's path, but have absolutely no chance to influence the path of a billion times billion heavier and greater (the most part of a golf ball is very far in this gravity field) golf balls.

But we can just as well simply say, that the forces by which are bound the surface layer of atoms of the edges to the layer of atoms underneath roughly equal to the forces in vicinity of edges at the distance approaching the size of atom so approaching the wavelength of light's or electron's dimension (wavelength).

The result of this influence of the large size of the gravitational forces on the microscopic quantum objects are curved paths in close proximity to the atoms of the surface of the material. The superposition of these curved paths around inner sides of edges with the recoiled paths from outer sides of edges is then observed as the constructive interference picture in the double-slit experiment.

In quantum mechanics, however, this natural difference between the behavior of macroscopic and microscopic objects, demonstrated by the double-slit thought experiment, becomes the basis for claims concerning the mysterious and beyond all understanding behavior of quantum particles. It allegedly is completely incompatible with the concepts of classical physics.

Johannes van der Waals (Nobel Prize in 1910) studied for decades, both experimentally and theoretically, the existence of mutual forces between the molecules and atoms of substances emerging as an averaging remnant (magnetic field) at their random thermal rotating movement of their thus rotating dipole and multipole electrostatic fields. The thermal averaging effect is much less pronounced for the attractive induction. Van der Waals also used the Greek letter Ψ for the free energy of a system with different phases in equilibrium at critical temperatures, describing the phenomena of condensation.

The Casimir effect is an experiment of the same nature as the double-slit experiment. The Casimir effect shows that the infinitesimal (non-measurable) forces of a small number of atoms of a material applied to macroscopic objects (golf balls) moving in the vicinity of these atoms (as is the case with double-slit experiment) may become an observable and measurable effect even for macroscopic objects if the number of interacting atoms increases many orders of magnitude, as it is in the case of the Casimir effect. A typical example is two uncharged neutral finite plates in a vacuum, placed a few nanometers apart at a distance comparable with the size of atoms.

As was shown in our previous paper [10, 11], at such small distances in classical descriptions, the gravitational forces (identical with magnetic forces) manifest themselves to a non-negligible extent.

But we can just as well simply say, that the forces by which are bound the surface layer of atoms of the plates to the layer of atoms underneath is roughly equal to the forces between plates if the distance between plates is approaching the size of atom.

A distorted physical premise was the basis of H. Casimir that, in a classical description, there is no field between the plates and so no force would be measured between them.

Casimir formulated the theory in 1948, predicting a force (Casimir–Polder force) between these plates on the mysterious claim that this force has nothing in common with plates and they exclusively flow from outside pressure of the vacuum, because not all wave lengths of simple harmonic oscillators of vacuums can fit between plates. According to the second quantization of quantum field theory, space is filled with zero vacuum point energy, containing an infinite quantity of oscillators of all possible energy values and wavelengths.

It is remarkable that, in quantum field theory, excitations of the field correspond to the elementary particles of particle physics. This is fully in contradiction with the Higgs boson theory that all particles in the universe have obtained their mass from Higgs bosons soon after the Big Bang.

Surprisingly, in the last decade mainstream physics came to state that Casimir effects can be formulated and Casimir forces can be computed without reference to zero-point energies. *"The Casimir force is simply the (relativistic, retarded) van der Waals force between the metal plates. They are relativistic, quantum forces between charges and currents. Thus it can be interpreted without any reference to the zero-point energy (vacuum energy) of quantum fields"* [22].

The consequence of this paper in the broader context should have resulted in immediate removal or at least suspension of Quantum mechanics in physics.

But the Casimir effect is dogmatically taught for fifty years at schools like evidence of correctness of QM approach for a truthful explanation of the physical reality based on non-measurable and physically unverifiable fantasies of frequencies and wavelengths.

7. Man's Ability to Perceive the Physical Reality of his Surrounding World

Exploring the rest of the world by man over the course of the last two hundred years shifts the world from what he can see with his own eyes, to the world of what cannot be seen with his own eyes. Man discovered that the functioning of the macro world of animate and inanimate nature, visible to him, is based on the functioning of an invisible micro world of cells, molecules, atoms, protons, electrons, and photons. He discovered that outside the cosmos of stars and the few nebulas visible to him with his own eyes there exist a space of galaxies and other structures in the universe.

If a physicist today sees a steel sphere about 1 kg in weight lying still on the table, he knows that inside, this ball is unimaginably "live". In this sphere of billions of movements and physical processes exist what physics is unable to grasp even at the level of atoms, or able to grasp only a small part.

Today we know that in this steel sphere there are a billion times billion the basic (roughly 10^{26}) construction elements of sphere-atoms of iron. These atoms are vibrating in their equilibrium positions, emitting thermal infrared radiation and electromagnetic fields. In its vicinity, tremendous gravitational and magnetic forces act (if gravity and magnetic force are not identical powers). Even greater electric power is located in each interior of these atoms of iron, where 26 electrons as force fields swirl in all various directions in 26 shells around the nuclei of these atoms. In each interior of these nuclei 26 protons and 26 neutrons most likely swirl in all various directions in force field shells. In each interior of these 26 protons and 26 neutrons swirl force fields as an unknown number of alleged gluons and alleged 3 quarks.

Physicists until 1919 had not even the slightest knowledge of this structure of the universe and atoms.

155

Also "live" is the vicinity of this steel ball. Today we know that in every surrounding cubic meter of space surrounding us (anywhere in the universe) there are more than a billion times a billion photons of electromagnetic radiation, at least a billion neutrinos, one hundred million photons of relict radiation, tens of hard actual particles of matter – protons and around a hundred thousand warm low density material atoms.

When looking at this steel ball from a distance of 1 meter, a physicist knows that (in line with his perspective) around 100 million atoms of air occur on this line. These atoms of air we don't see and we are not able to grab in our hands. When waving our hands in the air, we can sense the pressure of air resistance on our hands, evidencing the existence of fluid around us, in which we are plunged. This fluid air presses on the surface of our bodies with a force of around 15,000 kg and we do not feel this pressure if we find ourselves at rest in the air.

Convincing the general public of the veracity of the existence of this huge pressure required considerable effort from physicists. We recall the famous experiment of German scientist Otto von Guericke's in 1654 in which, after pumping air from the space of two half-meter hemispheres of iron, freely attached to each other via a seal, 8 pair of horses drawing ropes fixed to each hemisphere (a total of 32 horses) were unable to pull these hemispheres apart.

If the part of the air around us in a room is accidentally lit by the rays of the Sun we'll see a huge amount of dust in the air around us. We only see dust particles larger than roughly the size of 1 micrometer, so we see just particles containing more than one million times a million atoms. We can capture these aerosol particles by filters of nanometer size.

From the total amount of the entire electromagnetic spectrum of photons (wavelengths in range of more than 15 orders of magnitude from hundreds meter to $10^{-15} m$) which occur around us we can see only a tiny portion of less than one fiftieth of the whole spectrum (wavelengths in range less than one order of magnitude from 0.2 to 0.7 micrometers).

For creation of the optic perception in the eye the continuous stream of photons must last at least 1/16 of a second from one place. As the action of one photon with wavelength around 0.5 micrometers lasts $10^{-15}s$ we fail to see streams of photons of less than a million times a million. Therefore we don't see the spokes of a rotating bicycle or car wheel.

If, at a distance of 1 meter, a 1 cm thick and perfectly transparent glass sheet is placed before this steel ball, nearly 100 million atoms of glass will stand between, the existence of which our eyes provide no information. The existence of this sheet we learn only by its feel when we move our hand against it. Similarly, we hardly see living beings, such as transparent jellyfish, in the sea or stranded on the beach. We fail to see millions of viruses, bacteria and protozoa around us.

From a billion times a billion photons around us we see almost nothing or, we see just the photons that are important for our macroscopic life.

8. The Opposition of Physicists for the Past Hundred Years against the Deformed Physical Image of the World as Forced by Power Structures upon the Public

The history of mankind traces the fact that the more centralized the power structures of human society, the more the physical picture of the world and the freedom of human thought is under the control of these structures.

During that period in time (100 – 1600) when Ptolemy's geocentric mathematical description of the universe was under the supervision of the ruling power structures as an untouchable truth for 1500 years, is today described as a dark period in the history of mankind. Those were times when man was hindered in his progress in exploring the real physical picture of the surrounding world. The normal development of the physical sciences was stopped in its tracks for over 1500 years.

Next mainly the period within the 17th to 19th century is, once again, a clear example of a conspiracy of the power structures against the physical reality of the image of the world. In this period physicists rediscovered the discoveries of ancient Greek thinkers. The books of the most outstanding physicists (including Newton, Galileo, Tycho Brahe, Kepler and Copernicus) of this period (till 1835) were declared heretical and banned on Index Librorum Prohibitorum which was not abolished until 1966.

The ban also involved restrictions on printing these books in Europe. Most Greek natural science, other than Aristotle's with his geocentric universe was banned. Up to 1758 all books that supported heliocentrism were banned. Violation of this ban could lead to the death penalty. Not until 1992 was the Inquisition against Galileo repealed and was it admitted that the heliocentric approach in physics was correct.

In the 20th century, the crusade of the power structures against the actual reality of the physical world around us continues.

The experimental results of the research of whole generations of hundreds and hundreds of outstanding physicists in the field of mechanics (from Galileo and Newton to Mach) celestial mechanics (from Galileo to Hubble), electricity and magnetism (from Volta to Tesla) are again rejected in substantial parts in the 20th century and replaced by mystical theories of nonmaterial causation of metaphysics of time in mathematical structures of relativity and quantum mechanics of several theoretical physicists in the 20th century.

There are no unequivocal physical evidences for the mathematical construction of these theories such as special and general relativity, quantum mechanics, force fields theories using force-mediating particles, the standard model of quark theory, the Big Bang and the Higgs boson theories. These theories are based on mysterious claims of light velocity, of space time curvature, of non-existing of simultaneity contra existence of simultaneous body presences everywhere in the universe, of wave functions of bodies in infinity, of the invalidity of the law of conservation of energy.

After Edington's British expedition on Principe Island for the purpose of the observation a light bend of stars near the Sun at the Eclipse of the Sun in 1919, the greater part of mainstream physicists, by intervening, tried to prevent the publishing of Edington's articles. In spite of this intervention, bombastic subtitles arose in most mass journals, mainly in the UK and Germany and glorified relativity heavenward.

After this unsuccessful intervention, in 1920 the most respected physics of world, including W. Wien, P. Lenard, Sommerfeld, Nernst, Weyland, Debye and the Rubens based Union of German Natural Scientists organized a putsch in the Nauhaim Conference congress in 1920. Sommerfeld was the President of the German physical society in 1919-1920 and from 1917 until his death in 1951 he was each year proposed for a Nobel prize. Lenard was laureate of the Nobel Prize in

1905 and Wien in 1911. The content of the putsch was "*Einstein as a plagiarist; Anybody who supports the relativity theory is a propagandist; the theory itself was Dadaist (this word was actually uttered!)*"

"*Einstein's relativity principle could only achieve general validity by dreaming up suitable fields. The abolition of the ether was announced in Nauheim. Nobody laughed. I do not know if it would have been otherwise if the abolition of air had been announced*" Lenard retorted at one point [23, p. 239]. Lenard also argued that with his relativistic theory of gravity, Einstein had tacitly reintroduced the ether under the name "space". Lenard was a genius, operating with the deep conviction that only careful experimentation could advance the understanding of the structure of the universe.

Stark (Nobel Prize 1919), in the English journal Nature declared, "*The relativistic theories of Einstein constitute an obvious example of the dogmatic spirit,*" and he announced, "*I have directed my efforts against the damaging influence of Jews in German science, because I regard them as the chief exponents and propagandists of the dogmatic spirit.*"

Ernst Rutherford (Nobel Prize 1908) declared that the theory of relativity of Einstein, quite apart from its validity, cannot but be regarded as a magnificent work of art.

E. Gehrcke thought that relativity was a fraud and that its acceptance by the public was a case of mass suggestion.

P. Weyland believed that Einstein's theories had been excessively promoted in the Berlin press, which he imagined was dominated by Jews who were sympathetic to Einstein's cause for other than scientific reasons.

M. Planck's opinion was: "*To build a coherent theory, we must begin abandoning special relativity*"

A. Sommerfeld published in 1907 in Physikalische Zeitschrift paper 'An Objection Against the Theory of Relativity of Electrodynamics and its Removal'.

Ernst Mach (1838-1916) in his preface to the second posthumous edition of 'Principles of optics,' repudiated relativity as dogmatic.

A Michelson himself said that he don't like the theory that ensued form his work and was upset that his experimental work give birth to the oddity of relativity [49].

In a memorable confrontation at the first Solvay Conference in 1911, Poincaré asked Einstein, *"What mechanics are you using in your reasoning?"* and Einstein replied, *"No mechanics,"* which left Poincaré speechless. All that Einstein's formulation of relativity says by way of an explanation of length contraction and time dilation is that these phenomena are required to keep the speed of light constant. This failure of Einstein's theory to provide physical explanations for several of its basic assertions was what had led Sommerfeld to complain, with some justification, about *"unvisualizable dogmatics"* and *"the conceptually abstract style of Semites"*.

Relativity, contrary to standard physics, does not explain the physical phenomena in nature but prescripts, without any explanation of how this phenomena must be.

The attention needed applies mainly to the work executed by Michelson [24] in the Michelson-Gale experiment (1925). A massive interferometer experiment, spread over fifty acres outside of Chicago, detected a fringe shift of 0.236 of one fringe due to the Earth's rotation. This was in agreement with ether theory and within the limits of observational error.

The consequence of this experiment, in which Michelson himself corrected his previous result of Michelson-Morley experiment from 1887, considered then as negligibly small, upon which was built the whole relativity, in intrinsic science, should have resulted at immediate removal or at least suspension of relativity in physics. But political principles exceeded physical principles in the 20th century. This simply maintained the same state during the previous 2000 years, without major changes.

But most outstanding are Dayton Miller's experiments and papers (around 20) from 1905 till 1935 published in Nature, Physical Review, Science [47] (e.g. *"The absolute motion of the solar system and the orbital motion of the earth determined by the ether-drift experiment"*, Science, June 16, 1933, Vol. 77, No. 2007, pp.587-588, or in Nature, 1934, February 3, Vol. 133). Especially Miller's paper in Reviews of Modern Physics in 1933 details the positive results from over 30 years of experimental research into the question of ether-drift. It remains the most definitive body of work on the subject of light-beam interferometry.

Miller in his papers documented motion of the earth and its direction in the space determined by the ether-drift at velocities in the average around 200 km/s. Among other things, his paper lists: *"The fact that the result obtained by Michelson and Morley was not negligibly small was very fully set forth by Professor Hicks of University College, Sheffield, in 1902, in his important theoretical examination of the original experiment. The earth and the solar system are speeding through space at the rate of 125 miles a second or more, ten times the speed previously suspected. There is an ether drift, the earth does carry along with it through space some of the ether, whereas the Einstein theory was built upon the assumption and the results of the 1887 experiment that showed no such drift"*.

Everyone who puts effort and reads through this Miller 1933 paper will recognize how tendency and false is Wiki links https://en.wikipedia.org/wiki/Michelson-Morley_experiment.

Miller's works, published in all-world top tier physical journals, fulfilled the hundred year task of physicists for determining of the motion of Earth towards its surrounding luminiferous aether medium. This meant the final end of STR and GTR.

Miller had not received Nobel prize for his 30 years running experimental work about the ether drift but the Nobel prize received physicists (Einstein for his one heuristic paper in 1905 on photoelectric effect - deformed compilation of Lenard, Planck, Boltzman and Stark works- written at age 26 after two years work as patent office clerk, De Broglie for single one theoretical work in 1924 - his PhD thesis written at age 32, Schrödinger) theorists who in their mathematical structures concealed the existence of the ether and gradients of ether in space, around the Earth and around atoms.

The outstanding Miller explanation of all the possible physical causes, which it is necessary at interferometry of ether drift, take into account (unknown mutual directions of velocity of movements of bodies and the ether fields around them, several dynamic effects of light reflection in a point of reflection of interferometer), are after the violent introduction of STR concealed and replaced by a single non-material and nonphysical cause - the time dilation. The time dilation, however, is nothing else than a summary of all these possible physical causes transformed into the mystery of time.

For a period of one hundred years since the violent introduction of STR, instead of intense experimental exploration of these physical causes, thousands of physicists produced thousands of theoretical works about fictional original and new relativistic thought experiments. In these works, they allegedly discovered a new existing physical world which allegedly flows from confirming or disproving mutual mathematical mysteries of space and time in robust mathematical structures without regard to physical reality.

Miller's conclusion, already in 1925, was *"The effect (of ether-drift) has persisted throughout. After considering all the possible sources of error, there always remained a positive effect."*

Einstein's reply to Millers conclusion was *"My opinion about Miller's experiments is the following. **Should the positive result be confirmed, then the special theory of relativity and with it the general theory of relativity, in its current form, would be invalid**. Experimentum summus judex. Only the equivalence of inertia and gravitation would remain, however, they would have to lead to a significantly different theory."* [25].

Miller in his 1926 paper in Science Journal penned *"Either the Einstein theory must be modified to meet the new facts, or if such modification is impossible, it must be scrapped"*.

What power spawns that the relativity is at universities and schools all over the world taught for the next 90 years as valid and justified by the existence of fabulous null result of the Michelson Morley experiment in 1887 and on top of it development of experimental results in the next following 30 years is not mentioned and concealed to public and students?

From experiments in last decades we can point to Shtyrkov paper from 2005 '*Observation of ether drift in experiments with geostationary satellites*' [53] or the outstanding 15 year experimental and theoretical works of V.V. Demjanov concerning his M-M experiments resulting in more papers till 2013 [e.g. 58] determining the maximal measured ether wind velocity at 480km/s.

However, any experiments or any scientists are not able to change the decision of the power structures about the establishment of relativity as well as quantum mechanics as the only officially valid theories of the physical world because these theories are in accordance with the ideology of these structures. Even rejecting Relativity by Einstein himself so by its creator or rejecting Quantum Mechanics by Feynman who was awarded a Nobel Prize for Quantum mechanics are concealed and disregarded.

What is today's official picture of the physical world? We have allegedly a Universe created in a fantasy of Big Bang theory according to the creation principle. We have the fantasy of allegedly inseparable three quarks as holy trinity basic building blocks of nucleons. We have a fantasy of God's Higgs particle. It is allegedly proved by mathematics of STR and GTR that the gravity force that we feel when we hold things in our hands, or inertial forces of resistance that we feel when we accelerate bodies by our hands do not exist and are pure fictitious and pseudo forces.

Bodies in the Universe, around our Earth, particles around atoms, move without any material causation. Their movement is a consequence of God's will commanded by God's geometry (just as it was interpreted before Newton even in the time of Copernicus and Kepler) of nonmaterial fantasies of space-time mathematical structures of relativity and nonmaterial fantasies of wavelength–frequency mathematical structures of quantum mechanics. All parts of physics were allegedly 'improved' to metaphysics of space and time in four vectors mathematical structures.

An explanation of physical world physics of the 20th century is back in the medieval dark ages. Will this situation, fully distorting the real physical picture of the world, last 1500 years as it was in the case of Ptolemy's theory?

The procedure, under which Einstein first created the theory and then tried in experiments to demonstrate the validity of his theory, was typical for him and is typical for all main theories of the 20th century. If theories had heretofore represented a second step for the possible explanation of the experiment carried out in the first step, theories of the 20th century are produced in first step by theoretical physicists and in second step they look for experiments that explain these theories.

If the experiment was not in accordance with his theory, Einstein acted fully in accordance with the dictum which he proclaimed that if an experiment does not fit the theory, it is needed to change the experiment. This was the case in his proposed experiment (1915), gyromagnetic ratio, in which the value 1 should be measured by his theory. Einstein himself, in carrying out the experiment measured values of 1.02 and 1.45, but in transaction for the Physical society he reported 1.02 and discarded 1.45. Although experiments of other physicists during the next six years showed beyond a reasonable doubt that the correct value is 2, Einstein stubbornly insisted on his 1.02 value [26, p. 311].

Since the establishment of special and general relativity, quantum mechanics, force fields theories using force-mediating particles, the Big Bang and the Higgs boson theories to the present, hundreds of physicists and many associations of physicists around the world have shown fatal errors in these works and controversies concerning these theories.

A good overview of the physicists and the Association of physicists, though not complete, can be found in the publication of a 10-year long project [27], completed in 2006: G.O. Mueller- 95 Years of Criticism of the Special Theory of Relativity involving 3789 publications criticizing the theory. In chapter 2, *Catalogue of Errors for Both Theories of Relativity*, hundreds of fundamental errors revealed by tens of physicists are demonstrated and, as a result, both theories are doomed [http://www.kritik-relativitaetstheorie.de/Anhaenge/Kapitel2-englisch.pdf]

The message of the project to the German public is:

Since 1922 the criticism is suppressed, the critics are calumniated, the public is told lies about the scientific value of the theory of special relativity. In 1922 the physics community, as part of the greater science community, has broken away from the tradition of search for the truth, a rupture of the tradition - as far as we know - never before committed by a whole branch of science and with the knowledge and support of the greater scientific community.

We are confronted with the great mystery of modern physics:

(1) Why has the rupture of the tradition been tolerated by the whole "scientific community"?

(2) Why has it not been detected by the public?

(3) How can the academic physicists hope to continue forever without one day being called to account for their acting?

(4) What are the motives of the academic physicists?

During several years of research concerning the criticism of special relativity, we found the following answers.

(1) The public in Germany has been cheated since 1922 and is cheated by the influential scientific community until today. Academic physics exert strong pressure on newspapers, journals, publishers and congresses not to accept any criticism of special relativity for publishing. Critical papers are suppressed, critical persons are excluded from any participation in the scientific dialogue.

(2) The academic physicists believe that nobody can expose the truth about their actions to the public because the public would never dare to doubt the integrity of these scientists because of the great achievements of natural science in the last centuries, and that the general public will always trust the physics establishment more than any critics.

(3) The motives of the physics establishment are subject of several speculations. Probably the strongest motive is that physicists are thankful for a theory that "does not need the ether". This was the position expressed in Einstein's paper of 1905. But only 15 years later, in 1920 in a conference held in Leiden he discovered the need of an ether. The relativists were not amused about this conference of their master. This change of idea in 1920 should have led, as a logical consequence, to a revision of special relativity, which, however, has not taken place until today. This remarkable fact of non-revision seems to be a strong argument that the ether may be at least one fundamental motive.

About 1914, special relativity had already been directly refuted by several experiments and indirectly by the absence of experimental confirmation. The Michelson-Morley-Experiment and its repetitions have had positive results, in complete contrast to the relativist's propaganda until today of an alleged null-result: these experiments have found velocities of the Earth of about 6 km/sec (1887), 10 km/sec

(1902), 7,5 km/sec (1904) and 8,7 km/sec (1905), In 1913 Sagnac, with his rotating interferometer, also found moving fringes, the rate of motion of the fringes depending on the rate of rotation of his instrument. On the other hand, there were no experimental confirmations for the pretended length contraction and time dilatation. This desperate experimental state of affairs before World War I has never been recognized by the relativist textbooks.

The apparent great success of relativity came with observations of the Sun's eclipse in 1919 which were said to have confirmed the general theory of relativity. This supposed result was immediately rejected by several important critics in different countries as misleading the public (for instance: A. Fowler, Sir Joseph Larmor, Sir Oliver J. Lodge, H. F. Newall, Ludwik Silberstein in England; T. J. J. See in the USA; Ernst Gehrcke, Philipp Lenard in Germany). - But the relativists informed the printed media of that time about the greatest achievement of mankind! The public opinion was made enthusiastic about "Relativity" and was told that now both theories, the special and the general relativity, were undisputable truths and nothing less than a revolution of our thinking about space and time and gravitation.

In Germany critical authors are strictly outlawed since 1922 by academic physics and therefore unite the critical arguments against both relativities in a booklet titled "Hundert Autoren gegen Einstein" [A Hundred authors against Einstein] published in 1931, protesting against the "terror of the Einsteinians";

The critical reader comes to the conclusion that special relativity is an unreasonable theory propagated to the public in academic and high school teaching to be the greatest achievement, together with suppression of any criticism.

"Relativity" as a whole and especially "Special Relativity" as the first of two theories, are hailed by academic physics and their propagandists as one of the greatest achievements of mankind in the 20th century, as announced to humanity by the "new Kepler-Galilei-

Newton" and the like, revolutionizing our ideas of space and time. This picture of magnificence and glory can hardly be outdone.

The normally suspicious and critical reader of relativity textbooks and the original papers of Albert Einstein very soon find many points of the theory questionable and is irritated that no relativist author, not Albert Einstein himself nor his disciples, is ready or able to deal with these evident critical questions and irritations which arise from simple logical analysis.

An author who declares the same effect sometimes as "real" and sometimes as "apparent" (Einstein 1905) cannot escape the notion of what he is eventually going to tell. Instead, the relativists declare any criticism as incompetent and stupid and the critics to be maliciously motivated. Generally the relativists abhor common sense and advise the reader not to trust it, but they fail to show which better sense the brave relativist is using.

As a surprise in 1958 the Japanese Nobel laureate Hideki Yukawa is reported to have criticized Special Relativity in a conference at Geneva during the UN Conference on the Peaceful Uses of Atomic Energy: the difficulties are such that "it would probably be found necessary to have a breakdown of the special theory of relativity"

Significant is an abridgement from journal SOCIAL TEXT, November 2003, Center for the Critical Analysis of Contemporary Culture, Rutgers State University quoted in G.O. Muller project:

„The situation is rather bizarre: while in all other fields of science there is permanent criticism and discussion of theories, only in theoretical physics there has been organized the silence of a cemetery to protect relativity theory from critical arguments, and still more bizarre is that nobody in the sciences and the history of sciences seems to have noticed this cultural disaster. Since the special theory of relativity in the public opinion has survived until today as one of the greatest achievements of science, in reality it has been disproved since at least 1914 (experiment of Sagnac), and the few features of the theory which have a basis in reality (as example: $e=mc^2$) are no relativistic

effects and the famous formula is not the idea of Einstein: the relativists simply have usurped the achievement of Thomson 1881, Wien 1900, Poincaré 1900 and 1904, Kaufmann 1901-05, Hasenöhrl 1904 (see our documentation pages 131-132). - The ruins of the theory are covered by social constructions as the cult of genius and the powerful suppression of any criticism and even the critics as persons, and relativity theory thus has become a theory of simple socio-physics. The proofs and details of this diagnosis - physical and historical - you will find in our documentation. (...) ".

In addition to the publications and associations listed in the extensive research G.O. Muller project after 2006 until today, we can additionally specify the dissident physicists association mainly as thus:

The John Chappell Natural Philosophy Society - American Association of hundreds of physicists around the World (from 2014 the main branch of former Natural Philosophy Alliance). Since 1993 they organized conferences, published Conference proceedings and currently keep a database of around 6,000 publications of dissident physicists.

The database of publications of hundreds of dissident physicists since 2007 around the web archive vixra.org, that holds around 19,000 publications.

The General science journal, a web journal that from 2011 holds around 5,000 publications of dissident physicists.

Sankt Petersburg scientific association Sci-community - organizing since 1973 non mainstream physicists conferences and published up to date 37 volumes of Conference proceedings.

V. Atsjukovskij, member of more Russian scientific academics institutions, with his lifelong work from 1960 up to today's in topics of ether belongs to one of the most outstanding personalities of this association. Among many other works considerable is his comprehensive treatise form 2008 - General Etherodynamics [48] and his lifelong critical papers on relativity especially the last one from 2012 – Critical analyses of bases of the theory of relativity [49].

In this critical analysis, all experiments that allegedly confirmed STR and GTR are inspected and all alleged confirmations of STR and GTR by these experiments are refuted. All postulates (up to 10) of STR and GTR are dismissed and STR and GTR are rejected. In this paper the estimation of today's physics is - *"Physics has actually turned into a branch of mathematics, freely operating on abstract concepts — a multiplicity of dimensions of space, multiplicity of dimensions in time, a multiplicity of universes, all sorts of "curvatures" and "singularities" of space and time, as well as many other, which has no relation to real nature".*

"The main reason for the heavy state of science and, particularly fundamental science, is the exhaustion of ideas, emerged as early as at the beginning of the 19th century, which have replaced the materialist methodology by swank idealism, which put math above physics, elevated postulates, i.e. the hypothesis, boosted by "ingenious" scientists, in the rank of laws and actually proclaiming unknowability the world around us. Hasn't this at worldwide scale good organized fallacy just simply one reason consisting in the comfortable existence of all this worldwide scientific mafia?"

In this paper the following special official regulation of the USSR Academy of Sciences from the year 1964 is revealed:

" Any criticism of the theory of relativity has to be equated to the invention of Perpetuum mobile; to the authors their misconception has to be manifested; printed criticism of the theory of relativity must not be allowed. Because it is non-scientific".

V. Atsjukovskij, the member of more Russian academics scientific institutions during all his life as well as also today at his 87 is warning:

"Today there are no more reactionary and deceitful theories worldwide than Einstein's theory of relativity.

It is infertile and incapable of provide nothing to experimentalist who need to deal with arose tasks. Its followers are not ashamed of anything, including the application of administrative measures against his opponents. But the time given up by the history to this "theory" has expired. Barrier of relativism, erected towards the development of naturalistic science by concerned persons, is bursting under the pressure of facts and new applications, and it inevitably collapses."

"Einstein's theory of relativity is foredoomed and will be thrown out from science in the very near future."

Similar state regulations, disclosed in the Atsjukovskij paper, had always been replicated throughout all countries of the former Soviet east bloc. Thus there is no doubt that this regulation was in force in all these countries as Czechoslovakia, Poland, Hungary, Bulgaria, Romania, East Germany, North Korea, Albania, Mongolia and perhaps in that time also including China and Cuba.

So there is no doubt that acceptation of relativity was political decision and relativity is shielded by governmental power structures as a single one officially tolerable opinion about physical picture of world around us.

It is documented in this book that there is little doubt that identical regulation is to date adhered to by governments, academics, publishing houses and mass media all over the world, though never officially confirmed.

However, to the dissidents of physics after 1920 we have to include Einstein himself. As was detailed above in chapter 3, Einstein, in conclusions of his 1920 papers [1, V7, D31], [35] and 1924 [12, V14, D332] 'rediscovered' ether as the direct consequence of 'his' GTR field equations.

In these papers he declared that *"The ether has been resurrected in the theory of general relativity... according to the general theory of relativity space without ether is unthinkable... ..space and ether flows into each other.. we are not going to be able to dispense with the ether in theoretical physics so with continuum furnished with physical properties...**the theory of space (geometry) and time no longer represent intrinsic physics** propounded independently of mechanics and gravitation ....ether has to serve as medium for the effects of inertia ...every theory of contact action presupposes continuous fields, hence also the existence of an ether... According to our present conceptions the elementary particles of matter are also, in their essence, nothing else than condensations of the electromagnetic field... now it appears that ether will have to be regarded as a primary thing and that matter is derived from it."*

From all these declarations it is clear that these final Einstein reasoning replaced his previous reasoning in the mystery of time dilation in STR or in space time mystery in GTR and are discoveries of the highest importance for Einstein. But no top tier physical journal published these Einstein's discoveries of highest importance. No top-tier journals, newspapers or mass media celebrated and glorified to heaven on their front covers these ingenious new discoveries of the physical reality that now replaced previous metaphysical mysteries.

These new and highly important Einstein discoveries presented at more than 5 (from 1920 to 1934) conferences by Einstein (e.g. 1920 Leiden, 1924 Zurich, 1930 Berlin) fade away in proceedings and transactions from these conferences. These works are prevented from becoming known to the wider physical and general public. So, after 1916, the fate of this new Einstein discovery of the ether found exactly the same fate as works of all other regular dissenting physicists for the last 100 years.

As it always was in the past history of mankind, the physical picture of the world, mainly in the period from 1905 until the end of the Second

World War continuing to the present once again became the subject of a struggle between international power structures at the highest political level. Mainly from 1920, after the Congress in Bad Neuhaim to 1925 German physicists disagreeing with establishing relativity as a single description of the physical world, are isolated and persecuted. After the change of political power in Germany in 1933, the roles of the persecuted and the persecutors reversed and, after 1945, reversed again until today.

During the 20th century until today, physicists whose livelihood depends upon state power structures (the vast majority) and who dare to question or disagree with those power structures' established physical theories are admonished, isolated, persecuted or fired from jobs or cut from the sources of their livelihood.

The publication is impossible of physical views other than those officially adopted in journals and the mass information media. But it is imperative in science to doubt! An appraisal of physicists and other scientists (not only by principals but also for highest scientific awards) is executed on the basis of the number of papers and the rating of so called impact factors of the journals in which their papers were published.

The real fact is that the peer-review mechanism, by anonymous referees of these journals, declared as the patronage of the science, thrown down as mechanisms of inquisition censorship and has devastating consequences for the science. By these instruments, editors and publishers have decisive power over who will be a significant and award-winning physicist and what kind of physical image of the world is established within society.

This power of publishers in recent decades, in order to maximize their profits, attained an unlimited dictation and the terror of the publishers reigned against dissidents as well as all even conforming scientists. Scientists, for the publication of their many years work in journals not only fail to receive remuneration, but pay the full cost of

175

the edition of the magazine. Scientists are forced by publishers to waste considerable time to learn and write their articles in the format of complex polygraphic publishing typesetting systems like TEX or LATEX. Otherwise the chances of acceptance of their articles for publication are slim. Any grammatical error in the text is unacceptable and authors must pay for their own linguistic proofreading.

Publishing houses and mass media corporations have decisive influence on governments of states or vice versa, because the owners (media magnates) of these corporations are often persons participating in governments. Most of the printed physics journals in countries of the world are owned (or are in "cooperation" with) by just four transnational publishing houses.

We can point to Randy Schekman (born in 1948), a US biologist who won the Nobel Prize in medicine in 2013. He said his lab would no longer send research papers to the top-tier journals, Nature, Cell and Science. *"Leading academic journals are distorting the scientific process and represent a tyranny that must be broken. I have published in the big brands, including papers that won me a Nobel Prize. But, no longer. An impact factor was the toxic influence on science that introduced distortions. A paper can become highly cited because it is good science - or because it is eye-catching, provocative, or simply wrong.*

These journals aggressively curate their brands, in ways more conducive to selling subscriptions than to stimulating the most important research. In extreme cases, the lure of the luxury journal can encourage the cutting of corners, and contribute to the escalating number of papers that are retracted as flawed or fraudulent. I have now committed my lab to avoiding luxury journals, and I encourage others to do likewise."

Inspected top tier physical journals, the internet web archive arxiv.com, Wikipedia (fully false, misleading and demagogic articles about physical picture of the world, relativity and quantum mechanics in Wikipeadia) and by enormous financial resources supported massive production of demagogic documentaries are the main tools of demagogy and distortion of physics exploited by contemporary ideological power structures.

M. L. Corredoira and C. C. Perelman in their 250 page book from 2008, *"Against the Tide: A Critical Review by Scientists of How Physics and Astronomy Get Done"* itemize the testimonies of 15 notable physicists giving extensive descriptions of the mechanism of censorship, persecution, threats and blacklisting during last decades by top-tier journals, mass media and arxiv.com [55].

There are more than hundred 'scientific' institutions and universities (e.g. pontifical or ecclesiastical universities) around the world founded and ruled by ideological power structures. Hundreds of 'physicists' affiliated with or graduated from these establishments' influence on formatting of a 'desired' picture of the world as referees or directly as participants in editorial boards of physical journals and web archive arxiv.com. These are the 'physicists' who patiently, by misleading verbal constructions in Wikipedia, perseveringly prove the disputative correctness of shady Relativity or Quantum mechanics and in this way distort the physical picture of the world.

A flagrant example of such a 'physicist' is Robert A. Sungenis, Ph.D. who, in 2006, at age 51, earned a Ph.D. in religious studies from the Calamus International University (represented also in the UK and all over the world), an unaccredited distance-learning institution incorporated in the Republic of Vanuatu. In 2014, he produced a documentary, *The Principle*, with subtitle as a scientific documentary. Sungenis, in this 'scientific' documentary, together with support of physicists standardly featuring in dogmatic 'scientific' documentaries during last 15 years (L. Krauss, Michio Kaku, Max Tegmark, J.

177

Barbour, G. Ellis, K. Mulgrew) again rejects Copernican and Galileo's heliocentric models.

The documentary supports the long-discredited pseudoscientific principle of geocentrism in accordance with their religious beliefs. The heliocentric model in documentary is viewed as part of a conspiracy to undermine the authority of the church in society. In this scientific documentary it is again allegedly 'proved' that the flat earth is a center of the Universe where the Sun, all stars and galaxies rotate around the flat earth.

Around the world there is no known decisive reaction of national physical societies or academic institutions requesting denial of this documentary as a scientific documentary and as a heretic for science and requesting, eventually in court, the correction in the public statement of the authors in mass media that it is not a scientific documentary.

The inability of national physical societies or academic institutions around the world to defend the physical science in this and many other cases is, at the same time, an indication that these institutions can hardly be regarded as scientific institutions. They do not fulfill the role of bringing people a scientifically proven physical picture of the world. The general public is, by the silence of these dominant institutions, deliberately maintained in the confusion and chaos of countless so-called scientific theories.

The reason is that in most of these institutions academics and religious academics are interlocking and, in cases when they are not, they are afraid of state ideological power structures.

The main purpose of this and similar 'scientific' documentaries is creating chaos, confusion, fear and uncertainty in the public opinion on the physical picture of the world. The distinction between the true and a false (by propaganda enforced intentional lies) physical image of the world is, in the last two thousand years, even for professional physicists in this topic, the harder part of scientific work than the scientific work itself. The general public and also scientists not working in this area do have neither the will nor the time to carry out this hard work in distinction between the truth and a lie. They simply undertake what is forced on them by power structures through mass propaganda and endless repetition.

Einstein after 1920 fully rejected his void space-time conception in General relativity, rejected his STR, fully accepted that the mechanical ether must be considered by us as a physical reality and explicitly declared the inability to dispense with the ether in theoretical physics. But for the next 90 years just the space time conception from 1905 and 1915 are presented to the general public as allegedly Einstein's conception (which he later rejected).

So the situation with Einstein is the same as in Hubble's case. As was documented in this book, Hubble remained cautiously against the Big bang theory until the end of his life. In spite of this Hubble is by ideological power structures declared as the discoverer who proved the Big Bang theory. Einstein after 1920 admitted his fault from 1905 till 1915 and declared that his *"theory of space and time no longer represents intrinsic physics"*. So, although Einstein himself after 1920 rejected space time conceptions from 1905 and 1915, the next 90 years up to today these rejected conception from 1905 and 1915 are forced by mass suggestion of ideological power structures on general public as Einstein's ingenious new understanding of space - time and gravity as his ingenious conception of space - time curvature.

Of the hundreds of documentary films about the physical image of the universe produced in the last thirty years not a single one mentions the biggest experimental discovery of our civilization about the creation and the annihilation of particles from and into the electromagnetic radiation so from and into the ether.

All fictions and fantasies that are daily forced to the public as reality are also accepted by most of current civilization as physical reality. For the alleged confirmation of these science fictions are, without hesitation, lavished by enormous amounts of money in experiments at CERN or NASA.

The argument of academics for receiving money for building ELI, the largest laser facility in the world in the Czech Republic in 2015 was - *A team is planning to build an enormously powerful laser that could rip apart the fabric of space.*

A few years ago, NASA sent four satellites into space with gyroscopes to test the relativity theory, a project called Gravity Probe B. Just the fact that NASA is testing the theory speaks for itself. Why otherwise would you test something if it is right and taught for almost 100 hundred years as a reality in universities and secondary schools?

On the other hand, almost no money is granted to experiments searching for what constitutes the electromagnetic waves that we daily produce by our mobile phones (connection with operators or at Wifi or Bluetooth connection) or that we use for inductive charging of our mobiles without plugging them into an electric socket.

According to official GTR in this double situation the curvature of space and time flows into and out of our mobile phones. Instead of massive support for the finest experimental methods to explore the most subtle nature of the present real physical world around us, by enormous financial resources are supported the CERN big crash experiments, allegedly demonstrating how the physical world looked over 14 billion years ago or to demonstrate the fairy-tale of the rip of non-material time and space.

From 1905 until today there has been an opposition group of a hundred dissident physicists which consist, outside of admirable exceptions, physicists upon whom the ruling power structure have very small influence. These are physicists shortly before their retirement or in retirement, physicists who left physics and after migrating elsewhere to earn enough money, returned as independent physicists. It includes physicists who were fired from their jobs in physics.

These physicists do not hanker after glory, fame or awards. They have no reason to speak anything other than the truth. For no financial reward, they endeavor to fill the congenital need of human beings for knowledge of the actual real world around us.

In groups of the retired, or about to be retired, we found a huge number of professors, academics and scientists with remarkable scientific careers and even a few Nobel Prize winning physicists.

Of all the names of the celebrities of last decades, it is necessary mention the Nobel prize laureate in 2004, mainstreams theoretical physicist Frank Wilczek (born 1951), professor of physics at the Massachusetts Institute of Technology. At his 57, in his 2008 book - The Lightness of Being: Mass, Ether, and the Unification of Forces [60] – he claims that space and time is filled with primordial ingredient which has its weight and uniform density, which is the ultimate source of mass matter is built from and that the space and time is no mere empty and passive container. He claims that instead of classical word ether for this primordial ingredient he will use a new word Fabric, Texture or Grid. It is a dynamic Grid - a modern ether - which spontaneous activity creates and destroys particles. This new understanding of mass explains the puzzling feebleness of gravity, and a gorgeous unification of all the forces comes sharply into focus.

Here we see that a few brave mainstream scientists as e.g. F. Wilczek in his 57 or above-mentioned R. Schekman in his 65, after gaining highest professional awards, when they can no longer be heavily prosecuted by power structures for their pronouncement, are tired of the silence of the true state of science and consider as the necessary duty of their mission to finally tell the reality.

Of all the names of the celebrities of last century we have also recall Louis Essen Ph.D., Dr.Sc., FRS, O.B.E. (1908-1997). His research led to his development of the atomic reference quartz ring clock in 1938 and in 1955. In the 1960s and later, he was among the first candidates for a Nobel Prize. In defiance of this, in 1971 he published *The Special Theory of Relativity*: A Critical Analysis, questioning special relativity, which apparently was not appreciated by his employers. Essen, in 1978 said, "*No one has attempted to refute my arguments, but I was warned that if I persisted I was likely to spoil my career prospects*".

Though he could effectively work for at least the next ten years, he involuntarily retired in 1972 and died in 1997. In October of 1978 he published a paper titled *Relativity and Time Signals* in the Wireless world journal.

In this paper [28] he penned that *the comparison of distant clocks by radio is now a precise and well known technique. This was not the case in 1905, when Einstein published his famous paper on relativity and there is some excuse for the mistakes he made in the thought-experiments he described in order to determine the relative rates of two identical clocks in uniform relative motion. But there is no excuse for their repetition in current literature. The mistakes have been exposed in published criticisms of the theory, but the criticisms have been almost completely ignored; and the continued acceptance and teaching of relativity hinders the development of a rational extension of electromagnetic theory.*

These criticisms were rejected by Nature [the most prestigious journal in science]. ***It could be argued that the truth will eventually prevail, but history teaches us that when a false view of nature has become firmly established it may persist for decades or even centuries.*** *The general public is misled into believing that science is a mysterious subject which can be understood by only a few exceptionally gifted mathematicians.* ***Students are told that the theory must be accepted although they cannot expect to understand it. They are encouraged right at the beginning of their careers to forsake science in favour of dogma. Since the time of Einstein and of one of his most ardent supporters, Eddington, there has been a great increase in anti-rational thought and mysticism.***

From a recent author's we can recall Hans C. Ohanian. He has B.S. from UC Berkeley, Ph.D. from Princeton University, has taught at Rensselaer Polytechnic Institute, Union College, the University of Rome and today is Adjunct Physics Professor at the University of Vermont. From 1976 to 2008 he published more than half a dozen textbooks and several dozen articles on physics.

In his book, Einstein's Mistakes [26], hundreds of errors in all Einstein's publications from 1905 till 1943 are demonstrated (except for his 1905 paper on Brownian motion).

In his book, he further concludes that Einstein's unified theory was indeed an original, exclusive Einstein contribution-and it proved an unmitigated disaster. The most grandiose mistakes of Einstein's career were his several unified theories of electricity and gravitation. For nearly thirty years, from 1926 until his death in 1955, these were the central focus of his research. Guesswork (identical with works in special and general relativity) inspired by God and unsupported by fact are perhaps suitable for theology and theocracy but they are not suitable for physics.

Not surprisingly, all of Einstein's several attempts at unifying theories were trash, and it is the crowning tragedy of Einstein's scientific career that this was obvious to all his close colleagues. Out of our compassion and respect for the great old man, only a few could bring themselves to tell him so.

When the physicist Freeman Dyson arrived in 1947 at the Institute for Advanced Study in Princeton, where Einstein had spent his final years, he was eager to make contact with Einstein's "living legacy" and made an appointment with him. For discussion he got copies of Einstein's recent papers from a secretary. The next morning he realized that although he couldn't face Einstein and tell him that his work was junk, he couldn't not tell him either. So he skipped the appointment and spent the ensuing eight years before Einstein's death avoiding him.

Einstein's unified theories were a grand delusion. They led to papers and more papers on abstruse mathematics, but they never yielded anything of lasting interest in physics.

M. Born described in his paper [61] from 1955 the weak point in Einstein's work in those final years: *"Yet now he tried to do without any empirical facts, by pure thinking. He believed in the power of reason to guess the laws according to which God has built the world."*

As Edington used the same principle Born wrote in 1943 a paper disputing this Edington principle and sent its copy in his letter to Einstein. Einstein reply to Born's letter was: "Your thundering against the Hegelism is quite amusing, but I shall continue with my endeavours to guess God's ways."

In 1928, after Einstein announced another in a series of definitive solutions of his finally proven theories the hectic bombastic media and publicity furor from 1919 was repeated and this theory was a worldwide sensation. A thousand copies of the dry-as-dust journal of the Prussian Academy containing Einstein's paper were sold out instantly and several thousand extra copies had to be printed.

Eddington wrote to Einstein, *"You may be amused to hear that one of our great department stores in London (Selfridges) has posted your paper on its windows (the six pages pasted up side by side), so that passers-by can read it all through. Large crowds gather around to read it"*.

In the United States, The New York Times had anticipated Einstein's publication with the headlines " Einstein on verge of great discovery resents intrusion" and "Einstein reticent on new work; will not 'count unlaid eggs'". And when Einstein's paper appeared, the newspaper gushed, "The length of this work, written at the rate of half a page a year-is considered prodigious when it is considered that the original theory of relativity filled only three pages."

The New York Herald Tribune outdid the Times by printing a translation in its pages of Einstein's paper, including all those incomprehensible mathematical formulas. The Tribune had prearranged for the transmission of Einstein's paper from Berlin to New York via Telex, using a special code for the transmission of the mathematical formulas. Einstein contributed to the newspaper furor by offering a lengthy explanation of his new theory in the Sunday edition of the New York Times in which he called it the third stage in the development of relativity.

But the new theory was another dismal failure. Einstein had written down a set of equations that made no sense. After strong opposition from then most outstanding physicist, it took Einstein three years to recognize that his another theory was dead.

He included his final field equations for the unified theory in an appendix to the 1949 edition of one of his earlier books, The Meaning of Relativity, and The New York Times promptly reprinted the equations on its front page with the headline "New Einstein theory gives a master key to the universe."

This was wishful thinking, but the Times was just as stubborn as Einstein, and when the 1952 edition of the same book appeared, the Times greeted it again with the headline "Einstein offers new theory to unify law of the cosmos." The trouble with Einstein's "master key to the universe" was that it was not actually a key but only a dream about a key.

Einstein made so many mistakes in his scientific work it is hard to keep track of them. There were mistakes in each of the papers he produced in his miracle year 1905, except for the paper on Brownian motion.

And there were mistakes in dozens of the papers he produced in later years.

But there were in 1905 also a few similar papers on setting the dimension of atoms from Brownian motion [e.g. M. Smoluchowski, Bull. Intern. Acad. Crac. 1903, or works of the outstanding independent researcher William Sutherland in 1902 and 1904, e.g. The measurement of large molecular masses, 1904] Einstein worked on and did not cite these authors although he used their results in his work. That Einstein was in detail familiar with these works can be seen, e.g. from a memorial article he wrote on the occasion of M. Smoluchowski death in 1917 to the journal Die Naturwissenschaften recap the life professional works of M. Smoluchowski [12, V6, D48].

In 1938, Einstein in his book The Evolution of Physics claimed: *"physical concepts are free creations of the human mind and are not, however it may seem, uniquely determined by the physical world".*

As was shown in this book, Einstein all at once declared tens of weighty physical allegations and postulates, which simultaneously antagonized each other, without any deeper physical reasoning. Each single one of these terms, symbols and postulates had required extensive physical, philosophical and logical discourses even before their acceptation as the axioms of his physical theories and laws. In fact without these necessary elementals of how just the creator meant it he cannot be proved or disproved. All this extensive scientific discourses and extensive physical reasoning (inappropriate for genius bringing laws-breathed him by god) were imposed upon the following ordinary physicists in the 20th century. Consequently for the last hundred years physicist all over the world in thousands and thousands of theoretical publications exegete this "law" and quarrel about what is physical meaning and physical reality of relativity and quantum mechanics.

Expounding, confirming or disproving these theories, laws and commandments enunciated into non-physical mathematical structures and thus piling up math over another math was the task of the physicists of the 20th century, instead of exploring and proving the real material physical world. All about this, as was shown it this book, most of this allegations, postulates, axioms and heuristic laws Einstein after 1920 recalled. **Einstein's theories need not be disproved because with his declarations in his papers after 1920 Einstein refuted them himself.**

Just a familiarization of the general public about this fact is necessary.

Producing the physical law in Special relativity on the basis of Galilei's transformation, Einstein reversed the classical mechanics to Descartes physics of the 17th century based on mv, which was denied at the end of the 17th century. Later, all physics was developed to metaphysics of nonphysical quantity of time at four vectors metaphysics. With the help of de Broglie and Schrödinger, Einstein employing quantum mechanics, reverted quantum physics to Maupertuis metaphysics of the 18th century.

Classical physics is enunciated as incorrect, physics is disintegrated into many parts where everybody plays on his own background and can claim what he wants. Physical laws in this part are fully different whereas transition between these parts of physics are administered by so called approximations that overcome the different degree of power of physical laws in this different parts. This is precisely the status which is the intention of power ideological structures in the last two thousand years.

In this book it was shown that basic principles of reunification of these parts of physics is simple but it requires refusal of the Einstein system of energies and momentums which are the same quantities shifted just by a constant c.

Above detailed Einstein's declarations after 1920

"If we had based our considerations on the Galilei transformation we should not have obtained a contraction of the rod as a consequence of its motion. The theory of space (geometry) and time no longer represent intrinsic physics propounded independently of mechanics and gravitation. In Newton's theory of motion, space has physical reality- in contrast to the case of geometry and kinematics"

in closer context also means refusal of his energy momentum system.

After 1920 Einstein openly admits his younger thoughtlessness in his 25 to his 36 epoch from 1905 till 1916 when he formulated his relativity - *"My opinion in 1905 was that one should no longer talk about the ether in physics. But this judgment was too radical"*.

As was shown above, he fully rejected his physically incompetent approach based on the kinematics of the Galilei transformation on linear ratios of space and time. This means Einstein's confession of his non perception of basic physical principles of Dynamics of Nature so non-linear ratios of space and time (expressing change in densities of matter in space described by gradients of fields or by changes of the pressures or by accelerations or by changes of velocities).

Although Einstein after 1920 in his 40s fully understood the absurdity of the theories he had produced in the previous 15 years, for the next 35 years of his life he also produced no meaningful theory.

The reason for this professional disaster was that Einstein until the end of his life never understood the largest physical mistake of his scientific career which resides in his understanding of the unit of time as the independent physical quantity instead of a proper understanding of the unit of time in mechanics as the identical quantity with the quantity of the unit of velocity (detailed above in chapter 5.).

After 1920, Einstein in compliance with his claim that "*The theory of space (geometry) and time no longer represent intrinsic physics*" replaced space by the physical reality of the ether. But he never replaced time by the physical reality of movement or especially unit time by the physical reality of unit velocity as the scale of movement.

Instead of Einstein's mystery of space and time, physical realities are gradients of ether and velocities imparted to bodies by these gradients of ether.

But powerful ideological structures quickly seized new opportunities to return the physical picture of the world with the help of Einstein's incompetent mysterious theories of relativity from 1905 and 1915 back into line with their ideologies.

The 20th century continues the state of the destruction of physics during the previous 2000 years. This destruction was caused by installation of the Ptolemy geocentric image of universe consistent with the idea of the principles of creation as the only tolerable image of the universe for 1500 years till the 16th century. This destruction was caused during 17th, 18th and 19th century by a ban on the books of the most outstanding physicists as heretical (including Newton, Galileo, Tycho Brahe, Kepler and Copernicus) on Index Librorum Prohibitorum (along with the restrictions on printing in Europe).

But this ban becomes less and less effective from the second half of the 19th century, especially after democratic revolutions in Europe in 1848 and since civil rights became more and more a reality. So powerful ideological structures come up with the proven procedure at Ptolemy model and violently installed Einstein's incompetent mysterious theories of relativity from 1905 and 1915 as the only one tolerable image of the universe to return the physical picture of the world back into line with their ideologies. On top of that (as revenge to banned misbelievers) all those outstanding previously banned physicists, in addition including Maxwell, were refuted in substantial part by these theories of relativity. All previous physics was thus, in fact, impeached without looking for solutions in the physical reality instead of provided solutions in a mystery of space and time in relativity or frequencies and wave length in quantum mechanics. .

During the hundred years following 1915, generations of students of physics were brainwashed with relativity and quantum mechanics although these theories are not understandable to anyone, including the academics who lecture on them. An understanding of these theories is not even needed because these theories begin with the proclamation that the processes they describe are not understandable to sane sense and

189

their nature does not respect the basic formal logic of man. But, in fact the reality of these physical processes were not understandable just to a few theorists who created these theories or, in order to position themselves as geniuses, reverted the explanation of physics to mysteries and metaphysics allegedly understandable to nobody else but themselves.

As is documented above, academics who after more than 50 years of study, lecturing and experimental work are not able to understand the endlessly changed and corrected work of a 25 year-old patent clerk genius without sufficient education and experimental experience in physics. Academics who advocate and pretend an understanding of relativity and quantum mechanics simply reiterate drilled dogmatic mystical phrases. They are unable to answer the simplest questions about the tens of controversies and logical absurdities of these theories. They pronounce that the learned truth is implicated in mathematics.

These theories cannot be understandable from the point of physics a priory because they are based on mysteries of time dilation, space-time, de Broglie wavelength in infinity, fantasies about frequencies, physically undefined wave functions or energies and a great number of contradictory basic axioms. Quantities - time dilation, space-time, de Broglie wavelength, frequencies, wave functions - in relativity and quantum mechanics present these axioms as fact although they are non-measurable experimentally and unverifiable physically. By using contrived combinations and relations of these axioms as well as many other teleological parameters, mathematical operators or misleading approximations in robust mathematical construction, these theories come to results comparable with data measured experimentally. And so, allegedly these axioms are verified and represent physical reality.

For Galilei and Newton one equation F=mg (or exactly as the gradient of potential with defined parameter of this gradient as quadratic change of force) was enough to describe the gravity field in a generally understandable way. For the study of the General Theory of

Relativity, a man can study it all his life in tons of mathematical publications of tensor theories, differential and other geometry theories and tens of group theories.

At the end of this study man learns just one solid physical fact. General Relativity is valid because using fabricated teleological approximations (dismissed by Schwarzschild and many others [37, 38] as was detailed in chapter 3.) we receive a correct solution that equals the Galilei- Newton equation F=mg. The results of GTR are (allegedly) almost identical with Newton in weak fields and differ from him in very strong fields. After Einstein's declaration, it is pity that the correctness of GTR probably can never be tested because man can never test it in such a very strong gravity field.

Einstein's relativity will never be understood because, as was shown in this book, the creator did not understand it either and didn't know what he himself was factually talking about in STR and GTR theories.

A robust mathematical theory of relativity and quantum mechanics, in which contrived mystery parameters are combined in order to come to predetermined results cannot be understandable from a physical point of view.

The mass-media suggestion on the general public is performed with proclamations about Einstein's super geniality and his theories of space and time form 1905 and 1915, as the greatest achievement of human spirit in all the history of mankind. All of this despite the fact that, after 1920, Einstein himself rejected his own theories – *"**The theory of space (geometry) and time no longer represent intrinsic physics**"*.

The same mass-media suggestion is performed in proclamations of Schrödinger's equations (in fact Einstein's bases) of quantum mechanics as the second greatest achievement of human spirit in all history of mankind. This despite Feynman, the Nobel Prize laureate for quantum physics in 1966, proclaiming *"**I think I can safely say that nobody understands quantum mechanics. We have always had a great**"*

***deal of difficulty understanding the world view that quantum
mechanics represents*.**"

The question is - what is the sense to teach such theories at
universities, moreover as the most important and most time consuming
part of the physical education which even the creators of these theories,
themselves the most genial physicists and lecturers, do not understand?
The answer is – the sense is –the destruction of physics (in which no
consonant picture of the physical world exists and everybody can have
their own truth) in order to install and sustain predetermined dogmas of
ideological political power structures.

But the larger purpose is power and money. Political parties on
religious bases are still the most powerful and ruling political parties in
the countries of the world. Voters of these political parties vote for them
also for the reason that the ideology of these parties provided to voters
is allegedly confirmed as valid, because it is consonant with science.

This consonance of religion with science (Big bang, mystery of
space time, god's particles) is frequently and massively instilled to the
mind of these voters at every occasion by every possible manner. Vice
versa these ruling ideological political parties of countries support such
scientific projects which offer the consonant order of the universe with
their ideology.

Power ideological structures, not Einstein, are responsible for the
intentional destruction of physics in the 20th century. Power and
ideological structures of the world reached again great success. They
succeeded in again bringing physics from reality into fictive
metaphysical mathematical structures in which all movement in the
Universe has no material causation and is a consequence of God's will
as commanded by God's geometry. Einstein becomes the new Ptolemy
of the 20th century. Today nobody is taught Ptolemy epicycles and so it
will be with relativity and quantum mechanics. The question is not if,
but when. The nation whose government rejects relativity and quantum
mechanics from the curriculum at universities and thus stops

brainwashing their students by dogmas and mysteries will become in the next decades the superpower in a new scientific and technological revolution, as was in recent history the similar case of England in the 17th and 18th century.

Einstein, in around 1915, also defended Ptolemy's mathematical construction as valid because, according to Einstein, the mathematical description of moving planets from geocentric and heliocentric positions are equivalent and depend only on the choice of reference frames. This statement we hear even today from the mouth of ideological representatives of power structures. It is clear evidence how, for Einstein, physical reality was not his main goal and was of minor importance for him. Instead of bringing sharp differences between the wrong and correct physical approaches, he intentionally dimmed down the reality.

Orbiting planets (in Greek, the word planetai means wanderers) definitely do not move (wander) according Ptolemy epicycles so, on trajectories seen from the earth, where they once move in one direction and then stop and after making a loop (consequence that they orbits on various inclination to ecliptic plane) they move in the reverse direction, they appear to wander. Planets in no case stop in their orbits as it is described by Ptolemy's mathematical description.

On the bases of Ptolemy, mathematical description physics could never have discovered that around the great mass of celestial objects exist gravitational fields which rule motion of smaller mass objects located in them. On the bases of Ptolemy's mathematical description, physics could have not established the mass of earth, of the Sun and of all other celestial objects or have never established the motion of objects in other celestial constellations. The Sun can never orbit the Earth because the size of the mass of bodies always also means the size of their fixation in their own gravity field, thus in the gradient of ether surrounding them.

Einstein understood how to become the greatest physicist of the 20th century. His main goal was faked up mathematical theories which reject a material causation of movement of bodies in physics. Not correct physical theories, but theories required by political ideological power structures to provide him glory. It was not compulsory for him to defend his deformed theories, but the structures which chose this theory as valid were required to defend his theories. And today they still do.

Hegel's dialectic (G. W. F. Hegel 1770-1831), which was advocated by Einstein [61], is the tool that manipulates the consciousness of man into Hegelian conflicts of frenzied circular patterns of thought and action. The Hegelian dialectic is the framework for guiding our thoughts and actions into conflicts that lead us to grasp pre-determined solutions. The solution or result of this conflict is pre-determined before anything happens and no other solution is desirable. Reaction to this conflict is chaos and confusion that offers an opportunity to provide a pre-determined solution. Man remains locked into dialectical thinking and cannot see out of the box. Once we get what's really going on, we can cut the strings and move our lives in original directions outside the confines of this dialectical madness.

An example of Hegel's dialectic is the situation of the development of physics of the 20th century.

Although all physicists at the end of the 19th century came to full recognition that the existence of a gradient of ether is the material causation of movements of bodies in the Universe, Einstein's mathematical relativistic structures of non-material causation of movement of bodies in the Universe were violently installed in the physics of the 20th century.

Einstein's theories brought a conflict in physics by exploited obscure and physically complex physical experiments (M-M experiment, double-slit experiment) that resulted in chaos and confusion. Instead of appropriately correcting previous physical laws of physics he established faked up new laws of physics. He proclaimed all previous laws of physics as approximations of his new non-material space-time physical law.

On this manipulated apparent crisis of physics introduced in by Einstein he gives a predetermined solution **of new world order** of non-material causation of the physical world in his GTR equation $G_{\mu\nu} = \kappa T_{\mu\nu}$. In this equation G=T is the necessary beheld unity of basic freemasonic symbols G representing God's will, compass, geometry of great supreme architect of the Universe in unity with the second basic masonic symbol T representing gods temple, gods treasure, square and triangle. G and T, according Einstein in GTR, influence each other in unity and so, in joining these symbols, we have to see the David's star or the Eye of Providence.

Those who consider the last paragraph to be smiling conspiracy theory please listen carefully to the speech of the 35th president of the United States, J.F. Kennedy on the 21st of April 1961 (he was assassinated in 1963) about Secret Societies, for instance on YouTube – JFK - The Speech That Killed Him - from which we can highlight these extracted sentences: *"For we are opposed around the world by a monolithic and ruthless conspiracy that relies primarily on covet means for expanding its sphere of influence on infiltration instead of invasion, on subversion instead of elections, on intimidation instead of free choice, on guerrillas by night instead of armies by day. It is a system which has conscripted vast human and material resources into the building of a "tightly knit", highly efficient machine that combines military, diplomatic, intelligence, economic, scientific and political operations. Its preparations are concealed, not published. **Its dissenters are silenced, not praised.**"*

Looking at the current geopolitical situation around the world, we find that the validity of these Kennedy's words is fully filled out at the wars and power conflicts around the world as well as in the physical science.

Therefore, in order to step out of the box, out of the frenzied circular pattern of relativity, the time dilation and space time curvature as a metaphysical cause of changes of velocities has to be rejected by physicists right from the beginning. Right from the first, they must reject the second postulate of relativity that the speed of light in a vacuum is the same for all observers, regardless of their relative motion or of the motion of the light source. So relativity as a whole must be rejected from its beginning. Each physicist who fails to do so decisively from the beginning is trapped in a vicious circle of relativity (as well as of Quantum mechanics) from which there is no escape.

The theory of relativity has created a new form of thinking in which the traditional confidence to common sense and formal logic is unacceptable. Academics must give up their normal mental state, a healthy human intellect and basic principles of formal logic by being forced to lecture students, as well as the general public, exclusively in this officially permitted dogma.

For this reason, it is necessary from the beginning reject relativity and quantum mechanics according to which cause of the motions in the universe reside exclusively in the nonmaterial metaphysical causation of mutual ratios of the space and the time or frequencies and wavelength.

It is appropriate to mention that the participants of the 2nd International Conference on Problems of Space and Time in Natural Science, 1991, from the USSR, the USA, Canada, Italy, Great Britain, Germany, Brazil, Austria, Switzerland and Finland issued the following declaration:

Due to prohibiting or hushing up the publications which contradict Einstein's theory, modern theoretical physics and astrophysics have come to a crisis. We propose to give up teaching relativity theory in secondary schools, which would give time for studying the origin and development of classical methods in mechanics and physics. Teaching relativity theory in higher educational institutions ought to be accompanied by discussions of alternative approaches.

9. Conclusions. The Truthful Physical Picture of the Universe

The claim that the cause of gravity is the curvature of non-material void space and time is the greatest degeneration of physical and philosophical thinking in all the history of mankind.

The curvature of space in fact represents a change of spatial density of real material substance ether (dark energy in astrophysics, energy density of the vacuum in relativity, polarization of energy density of the vacuum in QM, zero-point energy fluctuation in QM, ether in electromagnetism) and the curvature of time represents the change of velocities imparted to bodies by this spatial change in density of the ether.

Time is a measure of the change of moving objects and time is not an arbitrarily chosen variable, which could by itself span, lapse and vary independently of the moving objects. Without moving objects, time does not exist.

In mechanics, units of time and units of velocity are firmly fixed and represent the same in inverse proportionality- basic comparative speeds of movement. The change of the time unit is the change of the basic comparative unit of velocity- change of the basic gauge of velocity

Space in itself is nothing more than a three-dimensional manifold, devoid of all form, which acquires a definite form only through the advent of the material content filling it and determining its metric relations.

Gravitational force as an individual fundamental power does not exist and gravitational force is not a universal fundamental attribute of matter. Such a huge force as is the force of gravity on the surface of the atoms (which equals to at least the magnetic force) was never found.

Gravitational forces and gravitational fields surrounding great ponderable objects are made up of the superposition of a huge number of disordered magnetic fields of atoms and elementary particles, from

197

which these mass objects are composed. Gravity is just a tiny remnant of huge electromagnetic forces placed inside these objects.

Elementary particles of mater are spin products of curl compressions of ether as a local increase in the density of the flatly distributed field of the ether in space. Thus is formed the mass of particles of matter (more likely 10^{11} neutrinos in proton).

As a consequence of this curl compression into spin particles, radial force fields arise circumambient to these particles as a change in density (radial gradient from particles surface to afar) of ether in particles surroundings. This force field (at the various situations called electrical, magnetic or gravitational forces) provides opposite pressure on the surfaces of particles [11] balancing the internal pressure of spinning particles that holds mass-energy inside and keeps particles together.

Inertial forces are forces of the resistance of particles and bodies emerging at accelerating particles and bodies against the force fields of ether medium surrounding them.

Acceleration leads to changes in velocities that result in pressure changes in the surrounding force fields on the surface of the particles of bodies that returns the change in the radius of the particles and volume of bodies.

The above mentioned physical concepts are the basis of the unification of classical mechanics, classical electrodynamics, special relativity, general relativity and quantum mechanics.

This concept allows us to explain [11] the physical mechanism between mass and sizes of the gravitational field in surrounding of ponderable objects, provides an opportunity to explain the origin of gravitational fields and inertial forces. We can thus explain the origin of the internal energy of particles of mass bodies, where the energy inside the particles of mass bodies comes from, how energy gets into the particles and how this huge energy in the particles is held. Special and General relativity, as well as Quantum mechanics, does without these basic physical considerations and premises, upon which such theories should essentially be based.

Ether constitutes real material substances, infilling the space in each unit of its volume. Ether is the real rippling and curling material substance (maybe curling oscillating neutrinos) with a vast amount of this curl in the form of electromagnetic radiation and large amounts of this curl in the form of mass particles, mainly in the form of protons. All the great mass objects are the concentration of these mass particles in small local parts of universal space. After the gravitational collapse of local shrinks and explosions of supernovae, this shrink of mass is spread again mainly by the electromagnetic radiation and neutrinos across the space of the universe.

Filling the space of the Universe with swirling ether is all that is needed for the self-evolution of the Universe as well as for creation of our Earth, Suns, Stars and clusters of Galaxies. If Darwin's scientific theory of the evolution of species is also valid at the beginning of living nature, then all our known universe's living and non-living nature formed itself in evolution, from the initial filling of the space of the Universe with swirling ether and without the outside need for the intervention of a supernatural creator.

It was shown in this work that relativity was not introduced into physics on the basis of a consensus of physicists. On the contrary, relativity was introduced to physics by the force of power structures through mass information media, despite the resistance and opposition of a majority of the time's most outstanding physicists. Today there are no more reactionary and deceitful theories worldwide than Einstein's theory of relativity.

It was documented in this book that relativity, quantum mechanics and Standard Model of elementary particles are fully inconsistent theories and have to be rejected from physics. It was documented in this work that Einstein's theories need not be refuted because, by his own declaration in his papers after 1920, he refuted them himself. Simply a familiarization of the general public with this fact is necessary.

It was documented in this work that, from sets of existing physically based theoretical explanations of several new physical phenomena of the 20th century, just those theories that are in conformity with the creation principle (but most distort the physical reality of world around us) are selected, supported and promoted by power structures and mass information media. All others are suppressed and repudiated. This crusade in prevention, hindering and distorting the independent development of physics in the 20th century in fact simply maintains the same state during the previous 2000 years, without major changes.

The testimony by all outstanding physicists in the period from the 16th century to the 19th century of the existence of the ether as the primary physical material causation of all physical processes was canceled and replaced in the 20th century by ideological power structures opting for the non-material causation of physical processes of Einstein's relativity and quantum mechanics. Despite that, in his 1924 paper, Einstein turned himself back into a more enthusiastic advocate of the existence of the ether than supporters of the ether before the year 1905.

Lies concerning the mystery of the non-material space-time continuum remain for a hundred years through the massive production of demagogic documentaries, inspected top tier physical journals, inquisitorial internet web arxiv.com and Wikipedia. Day by day these lies are forced upon both public and academics as the true picture of the Universe as it allegedly resulted from Einstein's theories.

The cult of personality of Ptolemy, whose incorrect mathematical constructions persisted as valid for 1600 years was replaced in the 20th century by the cult of personality of Einstein and his (later recalled) physically incorrect mathematical constructions of non-material causation of physical processes.

It was documented in this work that, for the last 400 years in physics, there has been no distinguished physicist (including Einstein after 1920) who would not have recognized that, without the existence of the ether, it is not possible to explain the physical world around us. Opinions of these physicists are massively distorted on a daily basis and falsified by mass information media and power structures that created the physical picture of world around us without ether.

This recognition of the existence of an actual physical media filling out the entire universe, from which all the particles of rigid bodies and force fields around these bodies in the universe are made up, is the greatest physical discovery of contemporary civilization.

But this discovery is concealed from us. This discovery is an actual physical source of amazement; however it does not fit the power structures any more than the rediscovery of the circulation of the earth around the Sun.

Therefore, instead of researching ether as the first task of our civilization, the development of physics was diverted into multiple physical mysteries and metaphysical mathematical constructions, introduced by the power structures through mass media.

The theories of the 20th century eliminate from physics:

- Inertial forces as the first and direct evidence of the ether. Inertial forces are termed fictitious or pseudo forces which do not arise from any physical interaction between two objects, but simply from the acceleration of the non-inertial reference frame itself. Also gravity is termed as a fictitious force.

- Accelerations, gradient of fields and forces (Einstein's relativity) because they directly point at changes of the density of ether.

- The existence of gradient of fields around atoms and particles (double-slit experiment, Schrödinger equation, de Broglie, Casimir effect) because they directly point at changes of the density of ether.

Accelerations, gradient of fields or forces are in relativity replaced by mystery of space and time quantities non-measurable directly experimentally. Gradients of fields around atoms are, in Quantum mechanics, replaced by fictitious mathematical construction of frequency, wavelength or wave functions non-measurable or are unverifiable experimentally.

It can be safely said that Quarks and Higgs boson do not exist. The waves of matter, wave functions, quantum entanglements and waves of zero point energy presented by the vision of quantum mechanics do not exist.

The claims that space and time is the fabric of a space-time continuum, that space can be ripped or torn, that time itself can slow down or be ripped, that there are possible parallel universes as well as wormholes, that black holes exist by the vision of current physics, that future events could precede and affect past ones are pure fantasy. They have nothing to do with physical science describing physical reality and belong exclusively to science fiction literature.

But all these fictions and fantasies that are daily swallowed by the public as reality are also taken by most of current civilization as physical reality.

The inability of national physical societies or academic institutions around the world to defend physical science is an indication that these institutions can hardly be regarded as scientific. They do not fulfill the role of bringing people a scientifically proven physical picture of the world.

All our known universe of the animate and inanimate nature of rigid bodies and force fields (also gravity) is the result of the interaction and the superposition of the electromagnetic force fields around atoms, molecules, and elementary particles.

The existence, interaction and superposition of electromagnetic fields are the actual physical basis for the explanation of new physical phenomena in the 20th century, rather than explaining these phenomena in mysteries of relativity and quantum mechanics.

Exploration of the subtle nature of ether and force fields around elementary particles is the main task of today's civilization. But instead of the massive support of the finest sensitive experimental methods to explore the most subtle nature of today physical world around us, the CERN big crash experiments persist. This allegedly demonstrates how the physical world looked over 14 billion years ago or demonstrates the fairy tale of the rip of non-material time and space and is supported by enormous financial resources.

If those financial resources and salaries which were lavished on these experiments and all other experiments proving the validity of relativity, quantum mechanics or creation of universe in accordance with the creation principle were spent for intrinsic physics, our civilization could today fly on anti-gravitational conveyors and dispose with non-limited energy resources.

But this is, unfortunately, also not an interest of the ruling power structures.

Let us hope that to revert the physics of the 20th century from its metaphysical way to a way of physical reality will not take 1500 years as was the case in the recent history of mankind.

References

[1] I. Newton, Mathematical Principles of Natural Philosophy, Daniel Adee, New York, 1846

[2] Carman, On the pin-and-slot device of the Antikythera mechanism, with a new application to the superior planets, Journal for the History of Astronomy 43, 2012

[3] Isaac Newton, Letters to Bentley, 1692/3

[4] Newton's Letter to Robert Boyle, On the Cosmic Ether of Space, Isaac Newton, Cambridge, Feb. 28, 1678-9

[5] Hermann Weyl, Space Time Matter, Methuen & co., London, 1922

[6] J. C. Maxwell, A treatise on electricity and magnetism, Calderon Press, Oxford, 1881

[7] E. Hubble, The observational approach to cosmology, Clarendon Press, Oxford, 1937

[8] A. K. T. Assis at. all, Hubble's Cosmology: From a Finite Expanding Universe to a Static Endless Universe, arXiv:0806.4481v2, 2011

[9] J. D. Bjorken, Theoretical ideas on high-energy inelastic electron-proton scattering, SLAC pub. 571, march 1969

[10] P. Sujak, Big Crash of Basic Concepts of Physics of the.20th Century?, Proceedings of the Natural Philosophy Alliance, 20th Annual Conf., Maryland, 2013, http://vixra.org/abs/1108.0017

[11] P. Sujak, On the General Reality of Gravity as Well as.Other Forces in Nature and the Creation of Material Particles and Force Fields in the Universe, Proceedings of the Natural Philosophy Alliance, 20th Annual Conf., Maryland, 2013,http://vixra.org/abs/1304.0046

[12] The Collected Papers of Albert Einstein, Princeton University Press, Open-access website

[13] T. Ferbel, The Top Quark and Other Tales of 'Discovery', Univ. of Rochester and Maryland, 2012, ww.thphys.uniheidelberg.de /~plehn/includes/ bad_honnef_12/ferbel.pdf

[14] CH. W. Lucas, Talk at Natural Philosophy Alliance Video conference, Open Debate on Did the Mainstream Discover the Higgs Boson, 2015

[15] V. Khachatryan et al., Evidence for Collective Multiparticle Correlations in p–Pb Collisions, Phys. Rev. Lett. 115, 2015

[16] H. A. Lorentz, Electromagnetic phenomena in a system moving with any velocity smaller than that of light, Proc. of the Royal Netherlands Academy of Arts and Sciences, Amsterdam, 1904

[17] Nikola Tesla, New York Herald Tribune, New York, 11. Sept 1932

[18] B. Hoffman, Albert Einstein: Creator and Rebel, Viking Press, New York, 1972

[19] B. Riemann, On the Hypotheses which lie at the Bases of Geometry, Nature, Vol. 8, Nos. 183, 184, 1998

[20] E. Schrödinger, Physical review, Vol.28, No.6, 1926

[21] A. Das, T. Ferbel, Introduction to nuclear and particle physics, World Scientific, London, 2003

[22] R. Jaffe, The Casimir Effect and the Quantum Vacuum, arXiv: hep-th/0503158v1, 2005

[23] M. Eckert, Arnold Sommerfeld, Springer, New York, 2013

[24] A. A. Michelson, H. Gale, and F. Pearson, The Effect of the Earth's Rotation on the Velocity of Light, Parts I and II, Astro-physical Journal, 61, 137-145, 1925

[25] Einstein, letter to Edwin E. Slosson, 8 July 1925

[26] H.C. Ohanian, Einstein mistakes, W.W. Norton and company, New York, London, 2008

[27] The G. O. Mueller Research Project, 95 Years of Criticism of the Special Theory of Relativity, http://www.ekkehard-friebe.de/, Germany, 2006

[28] L. Essen, Relativity and Time Signals, Wireless world,.Oct. 1978

[29] P. Sujak, On Energy and Momentum in Contemporary Physics, American Physical Society Meet., Denver, March 2014, Y33, 3

[30] P. Sujak, On gravity, other forces in nature and the creation of mass particles and force fields in the universe, Bulletin of the American Physical Society Meet., Savannah, April 2014, T1, 56

[31] E. Mach, The science of Mechanics, The open court publishing company, London, 1919

[32] Sujak P., 2016, Call to Repel the Physical Theories of the 20th Century, J Phys Math 7: 202, doi: 10.4172/2090-0902.1000202

[33] P. Sujak, Historical, Philosophical and Physical Reasons for Denial of the Main Physical Theories of the 20th Century, Proc. of the Annual International Congress on Fundamental problems of science and technology, July 2016, St. Petersburg

[34] P. Sujak, Call to Repel the Physical Theories of the 20th Century, Proceedings of the John Chappell Natural Philosophy Society, Second Annual Conference, July 2016, University of Maryland

[35] A. Einstein, Relativity, 1920, Henry Holt and Company, New York

[36] J. Renn , J. Stachel, Hilbert's Foundation of Physics,1999, Preprint

[37] A. A. Vankov, Einstein's Paper: "Explanation of the Perihelion Motion of Mercury from General Relativity Theory", The General Science Journal, August 2014

[38] A. A. Vankov, General Relativity Problem of Mercury's Perihelion Advance Revisited, 2010, arxiv.org/pdf/1008.1811.pdf

[39] A. Einstein, Spielen Gravitationsfelder im Aufbau der materiellen Elementarteilchen eine wesentliche Rolle?, Sitzungsberichte der Preussischen Akademie der Wissensch., 1919, pp. 349-356, p. 351

[40] De Sitter, A. Einstein, Proceedings of the national academy of sciences, Vol. 18, N 3, 15, March 1932

[41] G. Gamow, My World Line, Viking, New York, 1970, p. 44

[42] L. Kostro, Einstein and the Ether, Apeiron, Indiana Universty, 2000

[43] A. Einstein, Ether and the Theory of Relativity, Methuen & Co. Ltd, London, 1922

[44] W. Isaacson, Einstein: His life and universe, Simon & Schuster, New York, 2007

[45] K. Przibram, Letters on wave mechanics, Philosophical library, New York, 1967

[46] E. T. Whittaker, Theories of aether and electricity, Longmans, Green and co., Dublin, 1910

[47] D. Miller, The absolute motion of the solar system and the orbital motion of the earth determined by the ether -drift experiment, Science, Vol. 77, No. 2007, 1933

[48] V. Atsjukovskij, General Etherodynamics, Energoatomizdat, Moscow, 2008

[49] V. Atsjukovskij, Critical analyses of foundation of theory of relativity, Analititcheskij, obzor, Iss 2, Moscow, 2012

[50] J. Stachel, Einstein from B to Z, Birkhauser Besel, 2002

[51] E. Giannetto, The rise of Special Relativity: Henri Poincaré's works before Einstein, Atti del xviii congresso di storia della fisica e dell'astronomia, 1998

[52] Hicks W. M., On the Michelson-Morley Experiment Relating to the Drift of the Ether, Phil. Mag., V 3, 9–42, 1902

[53] Shtyrkov E.I., Observation of ether drift in experiments with geostationary satellites, Proc. of the NPA, 12th Annual Conf., Storrs CT, USA, 23–27 May 2005, V2, N1, 201–205

[54] Bjerknes, C. J., Albert Einstein, the Incorrigible Plagiarist, Downer's Grove, Illinois, 2002.

[55] M. L. Corredoira, C. C. Perelman, Against the Tide: A Critical Review by Scientists of How Physics and Astronomy Get Done, Universal-Publishers, 2008

[56] A. Unizicker, The Higgs Fake: How Particle Physicists Fooled the Nobel Committee, 2013

[57] M. Longair, Quantum concepts in physics, Cambridge University press, 2013

[58] V.V.Demjanov, Physical interpretation of the fringe shift measured on Michelson interferometer in optical media, Physical Letters A 374, 2010

[59] P. Lenard, On the photoelectric effect, Annalen der Physik, 8, 1902

[60] F. Wilczek, The Lightness of Being: Mass, Ether, and the Unification of Forces, Basic Books, 2008

[61] M. Born, Physics and relativity, International Relativity Conference in Berne, Switzerland, July, 1955

[62] J. Barbour, The End of Time: The Next Revolution in Physics, Oxford University Press, 1999

[63] H. Lorentz, Attempt of a Theory of Electrical and Optical Phenomena in Moving Bodies, E. J. Brill, Leiden, 1895

[64] K.F. Schaffner, Nineteenth-Century Aether Theories, Pergamon Press, 1972

[65] J. Renn, The Genesis of General Relativity, Vol. 1- 4, Springer, 2007

[66] J. Mehra, Einstein, Hilbert, and The Theory of Gravitation. Historical Origins of General Relativity Theory, D. Reidel Publishing Company, Boston, U.S. A,1974

[67] H. Minkowski, Space and Time, Minkowski's Papers on Relativity, Minkowski Institute Press, 2012

www.ingramcontent.com/pod-product-compliance
Lightning Source LLC
Chambersburg PA
CBHW032000170526
45157CB00002B/484